TALL TREES
AND
SMALL WOODS

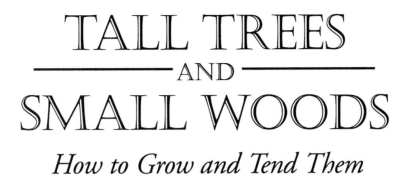

TALL TREES
AND
SMALL WOODS

How to Grow and Tend Them

WILLIAM MUTCH

MAINSTREAM
PUBLISHING

EDINBURGH AND LONDON

All line illustrations © William Mutch
Photographs © William Mutch unless otherwise stated

First published in Great Britain in 1998 by
MAINSTREAM PUBLISHING COMPANY (EDINBURGH) LTD
7 Albany Street
Edinburgh EH1 3UG

ISBN 1 84018 020 X

A catalogue record for this book is available from the British Library

Typeset in AGaramond
Printed and bound in Great Britain by Butler & Tanner Ltd, Frome

Contents

Foreword

The Scottish Forestry Trust's decision to support the publication of this book by Dr Mutch was based on the need to enhance and sustain rural economies by facilitating optimal use of land resources. An important component in this is to encourage forestry where it is environmentally and socially appropriate and especially if it can contribute to the rural economy. The right type of forestry should reflect local circumstances. This book by Dr Mutch is aimed primarily at farmers, environmentalists and people with a personal interest in trees and woodlands and will help to provide a better understanding of the establishment, silviculture and valuation of woods and their ecosystems, as well as the marketing of wood products.

The Scottish Forestry Trust is a charitable trust established in 1983 to promote education, training and research in forestry. 'Forestry' means the forestry industry as a whole, including its contributions to timber production, marketing of wood, landscape, recreation, wildlife and the environment.

This book fits into a pattern of Trust support for studies which highlight the need for more accessible information and better understanding of woodland management for people who are not fully familiar with all the technical aspects of forestry. A study, part sponsored by the Trust, into the socioeconomic factors influencing the decision to market timber pointed out the need for training in both technical and marketing skills for woodland owners. The authors of a report on the potential for extending forest cover in the Lowlands of England and Wales, which too was part funded by the Trust, called for farmers to learn about forest in order to ensure they have adequate technical knowledge to establish and maintain plantations. A subsequent study in the Grampian region, which examined the potential for extending forest cover in that area, indicated the advantages to be gained by farmers undertaking woodland work themselves rather than contracting it out. (Copies of these reports are available through the Scottish Forestry Trust.)

Dr Mutch's book conveniently gathers together a lot of information from various sources to provide guidance on the important phases of establishment and management of woodlands and on harvesting and marketing its production. In addition, a substantial part of the book focuses on environmental and conservation aspects which must be considered today in any form of land use, making this book useful to all people interested in managing woodlands for both conservation and recreation.

I am pleased to support the publication because it fills an important and unique niche in forestry literature. The information is presented in a form which is easily readable and understandable. Although other publications deal with much of the subject

7

matter covered here, they are often too detailed for the non-professional forester, at least initially, as well as often being difficult to find.

There is a need for more well-cared-for woods in the countryside, and I believe this book can help achieve that goal.

Sir David Landale, KCVO
Chairman, The Scottish Forestry Trust

Preface

This book has been written for anyone who is interested in trees and woods but is not a professional forester: for the farmer with small woods and hedgerow trees; for the multitude of community groups, in town and country, who are concerned about conserving trees in their area and want to plant a woodland, perhaps of native trees; for those who have trees in the garden; for those who have recognised that caring for woodland is one of the most exciting and rewarding leisure activities in modern living. Knowing more may give the confidence to begin taking effective care – and even if that is not do-it-yourself care, knowledge should help to define what a contractor should do and help to judge his performance.

The book is for people who would like to do more with trees and woodland but need guidance on how to get started. The lessons may be for everyone, whether they look after trees in some capacity or simply want to understand better what they see on a walk in the country.

The illustrations, especially those of the individual tree species, are really caricatures, not to be compared with the works of the serious botanical illustrator. In drawing them, I recall with thanks the teaching of the late Jack Anthony of the Royal Botanic Garden, Edinburgh, whose classes in Forest Botany in effect provided the basis for what is in chapter three.

At this time, when the trend in research and writing is towards greater specialisation, it may be surprising (and perhaps appear presumptuous) that one person, in one book of modest length, should attempt to survey the whole of woodland work, from the qualities of timber to the philosophy of sustainable management. With some reason this broad approach is out of fashion, since compression can lead to the distortion of over-simplification, while the provision of distracting detail lies as the opposing trap. Nevertheless, the fragmentation of the subject which specialisation is apt to cause appears to have led to the loss of appreciation of what forestry and woodland work involve and even to serious misunderstanding of them. The holistic review may help to remedy that loss.

Individual chapters of this book have been read in draft by friends – Baxter Cooper, the late Michael Leslie-Melville, Dr Douglas Malcolm, Dr David Rook, Dick Scruton and Dr Des Thompson – and I acknowledge their help with thanks. Each has made valued comments and suggestions which have greatly improved the text; nevertheless, the flaws and shortcomings remain my own.

I acknowledge my sincere thanks to the Scottish Forestry Trust for their generous financial contribution towards the costs of publication.

Finally my thanks go to my daughter Sheila MacNeill who has typed everything and to my wife Margot; without them the work would not have been completed.

W.E.S. Mutch
Edinburgh, 1998

9

Introduction 1

Trees and woodlands may be a personal pleasure, a local-community interest, the base of a national industry, an international duty – and all of these and more at the same time and place. At the end of the twentieth century we are in a renaissance of awareness of the importance of trees and forests, as people come to realise that these natural resources are an essential part of the sustainable production system of our planet. Trees support a huge complex of dependent organisms, absorb carbon dioxide, scrub atmospheric pollutants, give shelter and shade, provide a robust backdrop for recreation and, not least, produce a potentially never-ending supply of superbly versatile raw material – wood. Forest is one of the earth's great vegetation types, long treated as though it was inexhaustible, only now coming to be seen at its true value.

New woodlands are being planned and planted on many scales. Some are large schemes associated with the millennium or aimed at improving the fringes of major cities; others are the work of local conservation groups or of individual landowners. People are becoming more aware of the importance of trees in their environment and more woodland-conscious, valuing these habitats more highly, keen to extend and enhance them, ever more concerned at their destruction. This book may focus that awareness, improving the understanding and confidence of those who would like to become more actively involved and strengthening the effectiveness of their work.

Long ago almost all the lands of Europe, including the British Isles, were covered with natural forest – a wild wood – most of which has long since been destroyed. Most of it disappeared as a result of grazing, tree-cutting for timber and clearing to make farmland. As timber shortages became evident, landowners and eventually governments on an increasing scale began to create and manage woodland and forestry plantations as sources of fuel, carpentry and shipbuilding timbers and workwood of every kind. In the course of that planting, many beautiful woodlands and landscapes throughout Britain were created.

Britain's affluence as a trading nation in the nineteenth century made it feasible to supply timber requirements cheaply by importing the products of unsustainable exploitation in new lands. Interest in home woodlands then declined sharply, partly based on a belief that the rest of the world's forests were inexhaustible and partly with the displacement in shipbuilding of wood by iron and then steel.

Timber shortages during the First World War led to a sudden renaissance of home forestry and to a large afforestation programme. But there has been a common pattern in the recent history of Britain's woodlands and forests: economies of scale in wood processing have driven 'big-time' forestry to become ever more mechanised and ever more extensive, and at the same time the small woods and the small owners have been gradually left behind.

Professional interest in the care of small

woodlands steadily diminished over the last 50 years, accompanied by a general decline both in their productivity and in their appearance, which has been a concern not only to the woodland owners but to all who enjoy them as part of their environment. Small woodlands, when they are neglected in the presence of grazing animals or fire, are usually unable to regenerate themselves by natural seeding, as forests have done since the beginning of time. Too often the result of continued neglect is that they become unsustainable or perhaps are even deliberately destroyed as 'redundant'. We are all made poorer by their disappearance.

After decades of afforestation in Britain there are extensive forests, state and private, both on traditional estates and on the lands of forest management companies. It is an anomaly, however, that there is a weak tradition of farm forestry and almost no municipal woodland. There are few British examples of the culture in Bavaria or Scandinavia where farmers expect to draw a substantial proportion of their income from trees. Nor is there the equivalent of the communal forests of France or Switzerland, or of the town and city forests of Germany, Austria, Belgium or the Netherlands. A taxi-driver in Vienna will chat knowledgeably about sustained yield management and silvicultural systems, and citizens across Europe from Stavanger to Stuttgart are broadly aware of the current state of the wood markets, if only because sales of wood feature in the municipal accounts. There is a wide awareness of forestry in general and often a sense of ownership of 'their' woodlands.

In Britain woodlands are well regarded by the public, albeit in a rather vague way, but the absence of a strong forestry ethic means that public understanding of woodlands is weaker than in other countries. This is a handicap in any discussion of an integrated land policy and even in Parliamentary debate of forestry policy. The absence of a strong small-woodlands lobby in Britain is regrettable. Good steward-ship involves more than good intentions: it demands at least understanding of the concept of sustainable management, better still some practical experience.

An extensive, simply structured plantation forest, perhaps composed of a single, highly productive kind of tree, serves very well to provide the raw material for pulp and particle-board mills. Indeed uniformity, both in the kind of tree and in size, is necessary for selling wood to most bulk product manufacturers. Such material (and we all want the processed products of paper and panels to be as cheap as possible) has a low market value, which forces the professional forester to create and run a simple, tight system to yield as uniform a product as possible, because complexity in forest management is expensive. From personal choice and training, most foresters would prefer to work in woodland which is diverse and ecologically interesting, but market forces – which means the way we all spend and demand – require the bulk of our woody raw material to be cheap and uniform. As long as they are ecologically sustainable, such simple plantations are satisfactory and even commendable for their purpose. There are examples throughout the world: radiata pine in Australasia and Chile, eucalyptus in Africa and Asia, Sitka spruce in Ireland and Scotland, maritime pine in France.

That kind of extensive and necessarily simple forestry system, which is the response to market pressures for cheap raw material for bulk processing, is not at all appropriate for small woodlands. The natural forests, of the lowlands especially, and the managed woods which were developed from them were generally complex in structure with good variety in the kinds – that is the species – of trees. Where woods are expected to provide benefits and products other than the bulk supply of roundwood, other designs are possible, usually with several species, which are at the same time productive, attractive and ecologically robust. They can usually be managed to be more attractive financially than simple copies of the large commercial plantations.

Tending woodland and encouraging the trees to grow well is personally rewarding and highly creative work. It can make money, although there is generally far more made in processing wood than in growing it, but many of the results of tree-growing come as indirect benefits rather than as cash. The greatest American ecologist, Aldo Leopold, wrote in his classic book *A Sand Country Almanac*:

Every farm woodland, in addition to yielding lumber, fuel and posts, should provide its owner with a liberal education. This crop of wisdom never fails, but it is not always harvested.

What was true on Aldo Leopold's farm in Wisconsin still carries weight in all countries, not only with regard to woodlands on farms but also with regard to urban and community woods.

Throughout Europe the publicly owned forests, making up more than half the total area (some in central government and others in local authorities), are typically in large management units. In total area the private forests are not much less but they have quite a different structure, with a high proportion of very small woodlands, almost half being less than 20ha and a quarter less than 5ha. In the UK 60 per cent of privately owned woods are less than 20ha and 25 per cent less than 5ha, and Ireland is similar. Even these figures understate the number of small woodlands because in all surveys there is a cut-off point (say 0.25ha) and many small areas are not counted at all; nevertheless these groups of

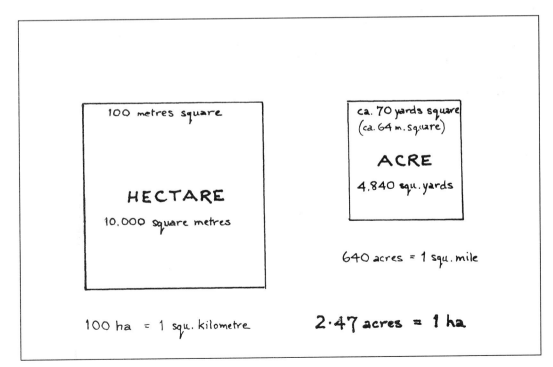

COMPARISON OF HECTARE AND ACRE

trees at field corners, on banks and along roadsides have great importance as landscape features and wildlife habitat.

Such fragmentation of the private woodlands presents problems for management. On small areas it may be quite uneconomic to use machinery, especially modern forest harvesters some of which can cut and process upwards of 100 trees an hour. Since bulk working is out of the question in small woods, different practices are needed, practices which are more flexible, more personal, more keyed to the soil and the local climate and more sensitive to local markets and to quality rather than quantity production. This is not merely an opportunity but a necessity. Small woods should not ape great forests; they can generally do another job better.

The attitude of many farmers to their woodlands is conditioned by their acknowledged lack of experience. In the British Isles, when the great landed estates existed, the farmland was rented to tenant farmers and the woodland was held by the landlord for timber production, game-shooting or both. There was no call for farmers to have expertise in woodland work, and so it has persisted to the present, even though the estates have been broken up and the farms are now owner-occupied. A farmer may admit the lack of attention, even say he has never thought about working the woods which are now part of the farm, as though he expects to be exploited by a smart contractor and decides to avoid trouble by doing nothing at all. The result is certain to be negative: either the woodland deteriorates by neglect or it is indeed badly handled by a contractor whom the farmer cannot direct, thus confirming the view that woodlands are all bother and best avoided. He should think again.

Currently it is public policy to create new wooded areas to serve the needs of townspeople – countryside-around-towns projects, community woodlands, country parks and so on. Around many cities there are also existing woodlands, most quite small, serving a multitude of purposes as play areas, dumps for rubbish, dog exercise grounds, adventure playgrounds, sites for bonfires or places to gather fruit or pick flowers. All these woods, even without help, also give living space to a huge variety of creatures which crawl and fly: birds, hedgehogs, squirrels, foxes, badgers and even roe deer which have learned to live with us, sometimes so unobtrusively that we may not know they are there, as well as the plants and insects which sustain them. For every kind of reason, these small woodland areas deserve our care so that they thrive and continue to serve us in the multitude of sustainable ways they can.

Throughout this book attention centres on forest trees rather than on ornamentals and shrubs, but little or no reference will be found to powerful machinery or the techniques of bulk-production forestry. Big machines are generally out of place in small woodland and people will find that hand-working with the tools and many of the techniques which preceded mechanisation are not only possible but more appropriate. In no sense are the techniques of do-it-yourself forestry second-rate, rather the reverse – careful work allows high-quality trees to be produced for valuable markets, with added benefits for the worker, for the environment and for other creatures.

Real values have not greatly changed in 2,000 years, since the Roman poet Horace wrote:

This was in my prayers: to have a parcel
of land not too large,
With a garden there and a spring of fresh
water by the house,
And, with these, a piece of woodland.

How Trees Grow: The Basis of Woodland Ecology

2

An arable farmer has the opportunity each year, if he wishes, to alter the soil to match the needs of the crop he is to grow, by cultivation, fertilising, modifying the acidity by liming and so on. He may grow a crop which defies the local exposure by using the shelter of hedges, or by planting a quick-maturing variety, or even by reckoning that the loss sustained in the occasional disastrous year will be more than offset by the successes of the good ones.

The forester, however, is denied these options. Cultivation is possible only before planting the trees, and the opportunity for effective change of the chemical condition of the soil is very restricted, perhaps only to tide the tree over an initial nutritional deficiency. Furthermore, trees must take the rough of the local climate with the smooth over the whole of their long life, 50 years or a century or more. The forester cannot take the risk of a heavy frost which may come 'only' one year in ten; when it happens his woodland of too-frost-tender trees will be dead and there will be no chance, as the farm has, of offsetting the loss against the nine good years.

By their very nature, trees stand tall and in the wind. There is no hedge for them to shelter behind, at least beyond their seedling stage. Woods must be designed and managed with careful regard to the local soil and climate. These may never be disregarded.

The assemblage of plants and animals which develops in any part of the world is not a matter of chance but a reflection of the soil and local climate, and the soil is itself a result of the action, over a long period of time, of weather and climate on the rocks. In each of the broad divisions of the plant and animal community across the globe – tropical rain forest, desert, temperate deciduous forest, tundra and so on – there is a typical set of plants, including trees, which are truly at home there and would be out of place elsewhere. For instance, in the boreal forest or northern coniferous zone in Scotland, the Scots pine and birch are typical but the beech and lime are not, these two belonging naturally to the temperate deciduous forest in southern England and most of western Europe.

In some places the natural set of plants has been restricted by an accident of the past. The fact that the northern coniferous zone in Britain has only two conifers, Scots pine and juniper (and the latter is scarcely a tree), is due to the flooding of the English Channel after the last ice age, separating Britain from the continent before spruce, larch and silver fir had time to spread north again from southern Europe as the post-glacial climate improved. In such instances it may be useful to bring in trees from elsewhere which will grow well, but introductions involve a degree of risk because of the trees' susceptibility to early or late frosts, changed day length, wind or some other climatic factor different from their home. There have been great successes among tree introductions to Britain and elsewhere in the world, but the use of native species and the local strains (or provenances) of them carries a good assurance of avoiding risk and gaining wider ecological advantages. The evidence of which trees are truly native is worth careful note.

The physical factors of climate, topography and soil limit the growth of trees and restrict how woodland may be managed.

CLIMATE

The climate is the combination of factors which exerts most control on woodland in Britain, imposing least constraint in the south-east of England and progressively more as one moves from Kent north-westwards to the Hebrides. In Britain the two principal components are accumulated temperature and 'exposure'. The warmth factor is measured by the accumulation of day-degrees above 5.6°C (42°F), which is the critical minimum temperature for plant growth. Below that temperature most plants do not grow, above it they do, so the aggregate count of the days and the degrees of warmth above 5.6° is a useful way of comparing the growing season of different places.

The key factor in the composite term 'exposure' is the mean annual wind speed. Rainfall is less of a constraint for tree growth in Britain than wind and warmth, because generally rain is sufficient throughout the country, provided the soil has reasonable capacity for storing the moisture it receives. In countries with less even distribution of rainfall (or snowfall), this generalisation does not apply, and instead other climatic factors, such as summer drought, may act as constraints.

TOPOGRAPHY

Topography can provide shelter for some areas and leave peaks and ridge crests severely exposed. Landform strongly affects the occurrence of frost damage to trees, with cold air flowing down slopes and accumulating in 'frost-hollows'; this has a major influence on the choice of species. Slope alone may control where wood may be harvested and where it may not.

SOIL

The depth to which trees can root is a critical factor in their growth and stability. It may be the depth of soil over rock or an impermeable

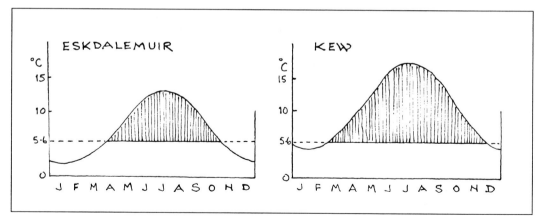

DAY-DEGREES

This is a way of comparing climates. The shaded area above the line 5.6° C represents the cumulative warmth at the site; the length of the base of the shaded area is a measure of the length of the potential growing season at Eskdalemuir and at Kew.

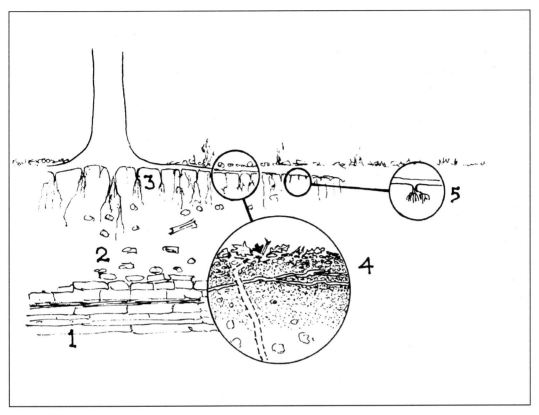

SOILS AND ROOTS

1. Bedrock is the parent material of all soil, whether it accumulates in place (2) or is carried by ice, water or wind to be deposited elsewhere.

3. From its main lateral roots the tree puts down sinker roots, anchoring itself and drawing in moisture and nutrients from the lower soil.

4. Forest litter accumulates on the soil surface, leaves and dead twigs: bacteria, fungi and mites decompose them into humus which is mixed into the soil mainly by earthworms. Many trees develop feeding roots close below the litter layer where they are vulnerable to damage. Earthworm tunnels provide soil aeration.

5. On gley soils the tree's sinker roots are annually killed back by the seasonal rise of the water table, so that they resemble shaving brushes.

layer such as compacted clay or an ironpan, or it may be the depth down to a layer of waterlogging.

Shallow soils present special difficulties. Shortage of rootable volume for trees tends to make them slow-growing and relatively unproductive. Clay soils may be easily damaged by motor transport compacting them or cutting them up.

The natural availability of soil nutrients is a key factor in determining the species of trees which will grow successfully on a site. The principal elements required, as for all plant growth, are nitrogen, phosphorus and potassium, although a few trace elements are sometimes critical. In most mineral soils these major elements may be sufficient for tree growth, but on those derived from acid rocks such as quartzite they are often deficient and require supplementing, frequently with phosphorus at the time of planting. Peats, apart from those receiving water draining from neighbouring mineral soils, tend to be nutrient-poor, requiring fertiliser application when trees are planted (and often also in later life, when much of the nutrient capital may be 'locked-up' in the wood and foliage of the trees themselves).

Waterlogging can be a complete barrier to roots and the cause of poor growth of the whole tree, due to poor soil aeration and low soil temperatures. In effect it restricts the volume of soil the tree can root into; not only does the tree grow slowly but its roots lack weight and it is more easily overturned by the wind in exposed areas. On some gley soils (clays often with blue and yellow streaks), tree roots typically show 'shaving-brush' growth, the root tips repeatedly trying to grow down in the summer when the water table falls and being forced to die back when the table rises again each winter. Sites which regularly flood are unsuitable for normal afforestation and the choice of tree species to plant is very restricted. Roots need air.

Droughtiness occurs in soils which drain freely and have a poor capacity for holding water. On sand dunes and old shingle beds the choice of species for planting will be restricted.

WINDTHROW

The British Isles have a much windier climate than the neighbouring Continent and wind is a major factor controlling forest management, especially in the uplands of Scotland and on the west coast. Windthrow is likely to occur where the soil is shallow and exposure severe; its risk is increased by some kinds of ploughing before planting and by thinning a plantation later than normal. Professional foresters assess the windthrow risk on a point-scoring system which takes account of the wind zone, the elevation, the topography and the soil type, thus providing a windthrow hazard class, on a scale of one to six. In classes five and six the risk of severe wind damage has a major effect on the species of trees to be planted and on silvicultural practice.

SOIL TYPES FOR FORESTRY

Tree growth and stability depend heavily on root development, so the physical attributes of the soil such as the rootable depth, aeration and the presence of compacted or cemented layers are the prime interest. Look at soils carefully and recognise their controlling influences.

Brown earths

These are free-draining fertile loams, usually slightly acid, which allow good rooting. Leaf-litter breaks down into brown humus which is well mixed through the depth of the soil, generally by the action of earthworms, so they look like good garden soils. A wide range of trees grows well but weed growth may be a problem at planting time.

Podzols

These are free-draining sandy soils, strongly acid and less fertile than the brown earths. Leaf-litter, typically from heather, breaks down to a greasy black humus (a 'mor')

WIND ZONES IN GREAT BRITAIN

Wind greatly influences tree growth and woodland management. The highest average wind zones 11 and 12 are in the north and west, the lowest windiness in south-east England. Comparable data for Ireland not available. (Reproduced by kind permission of the Forestry Commission, based on Research Information Note 230 by C.P. Quine and I.M.S. White, 1993)

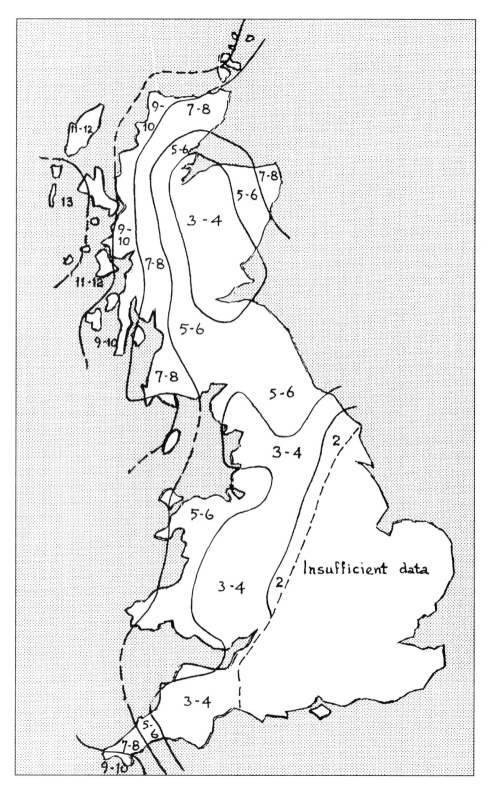

which is poorly mixed with the soil. Immediately below the humus layer the soil is bleached white, and deeper there is usually a deposit of dark staining where iron or other metals are deposited, having been leached chemically from the upper layers. Some acid brown earths may be slightly podzolised, with a layer of bleached soil particles.

Ironpan soils

They are usually found in the uplands and are similar to the podzols in the mor humus and acidity, but with a thin cemented ironpan which seriously interferes with drainage and rooting. They are infertile but can be readily improved by cultivating to break the pan (for instance by tine ploughing) and perhaps by planting birch, the roots of which have the power to penetrate slowly. Heather and purple moor grass are typical.

Gley soils

These are mineral soils with a high and perhaps fluctuating water table, often showing streaks of yellow and blue colouring. *Lowland gleys* are heavy clay, usually limy and fertile, but often liable to dry and crack in summer. In the wetter winter conditions the water table rises and poor aeration kills many roots which have begun to grow down in the previous summer. This produces 'shaving-brush' root development, except on specially adapted trees, such as alder. Tufted hair grass is typical. *Upland non-peaty gleys* and *peaty gleys* are similar, except that the water table is kept high all through the year by high rainfall which also causes shallow peat to develop (up to 45cm) on colder sites, usually with purple moor grass and either cross-leaved heath or heather.

Peat

Where dead vegetation does not decompose because conditions are too wet, too cold or both, the plant litter accumulates to form peat which may develop to many metres in thickness. Peat grown in water draining from higher ground may be reasonably fertile because the water brings nutrients from the soil it has flowed over or through. In contrast, peat supplied only with rainwater is very infertile. The former, flushed peat, typically grows rushes and purple moor grass, whereas the rainwater peat, on raised and blanket bogs, grows deer sedge, cross-leaved heath, cotton grass, sphagnum mosses and heather.

Calcareous soils

These occur over chalk and limestone rocks, mostly in southern England and the Pennines. Because of the high lime content, the shallow versions of these soils are difficult for tree-growing, but when they are deeper they are generally brown earths and very fertile.

Descriptions of soils (and acid rain, stream management and pollution) may refer to 'pH'. This is a scientific measure of acidity. Soil with a pH of 6.5 is approximately neutral – neither acid nor alkaline. Lower numbers are increasingly acid (it is a logarithmic scale); higher numbers are increasingly alkaline. A pH of 4 would be typical of an acid podzol with heather in an area of granite rocks; water from a severe 'acid rain' incident might be pH 3, which would kill fish. In contrast the soil in an ash woodland with dog's mercury over soft limestone rocks might show pH 8. There is no need, however, to measure the pH of soils for practical woodland work; the evidence of the plants growing on an area is quite sufficient to guide the choice of suitable trees to plant.

A crucial development in British and Irish forestry 50 years ago was the introduction of ploughs drawn by crawler tractors to prepare land for large-scale afforestation. This allowed wet land to be drained, ironpan and compacted layers to be broken up with deep tine ploughs and, with huge effect on successful establishment, a hospitable site for each tree to be prepared, giving local drainage, temporary freedom from weeds and local shelter. Such work allows some of the site constraints to be overcome, especially those concerned with waterlogging, soil aeration and the rootable volume of the soil.

Trees have an advantage over other plants because they are tall, dominating their surroundings. The stiff woody stem supports a crown of leaves which intercept sunlight above the competitors; the larger the stem, the greater the crown it can support and the more sunlight the tree can trap. How each structure functions is a guide to good silviculture.

ROOTS

Tree roots have three functions: to anchor the tree to the ground, to absorb water and to take in mineral nutrients for growth. The watery solutions absorbed from the soil are drawn up the tree in the woody tissue, the *xylem*. In order to work, the root requires carbohydrate food, which is passed down from the leaves through the inner bark (the bast or *phloem*), and oxygen, which it takes in from the soil atmosphere. Hence the need, in all but specialist trees, for the soil to be reasonably well drained; in waterlogged soil there is no atmosphere, no oxygen and the roots die.

The rootable volume of the soil, the depth multiplied by the area over which the tree's roots can extend, is critical for the tree's success. If the rootable depth is shallow and the individual tree has little space because it is hemmed in by its neighbours, it has limited raw material for feeding and little weight of soil in its roots to ballast it, a flaw which may lead to fatal instability as it pushes its crown up into the wind.

The mineral elements required for growth make up only a tiny proportion of the growing tree, perhaps five parts per thousand. Trees operate a highly efficient recycling system: the discarded leaves, fallen twigs and dying fine roots (which are easily overlooked in the system because they are out of sight but which have to be constantly regrown) are all broken down by a chain of bacteria, fungi, soil mites and earthworms, thereby supplying the tree with most of its essential elements. If only the timber is removed at harvest, leaving on-site the twigs and leaves which contain most of the absorbed minerals, hardly any scarce material is lost and, given the usual rate of annual deposit in the form of dust and minerals dissolved in rainwater, the site will be in healthy positive balance over the tree's life. In this condition, the typical one, the forest is a soil-improving system.

Nitrogen, in a form available to plants, is one element which is liable to be in short supply in some tree-planting areas. Some trees, notably the alders, have a special relationship with certain bacteria which can fix nitrogen from the air and form nodules on the feeding roots (bright orange on common alder); by this arrangement the alder obtains nitrogen. Other species benefit from the ability of many fungi to break down organic matter such as leaf-litter directly into available nitrogen compounds; the associated fungal structures on the tree's roots are called *mycorrhizas*. The growth of the tree may be virtually dependent on the success of the mycorrhizas; if the fungus does not thrive, the tree cannot grow.

The point at soil level where the root and the shoot meet is the *collar* of the plant.

SHOOTS

In spring each leaf-bud on the shoot develops into a leaf-stalk (the *petiole*) and a blade which is broad and flat in broadleaved trees. These broadleaves are called *hardwoods*, irrespective of the actual hardness of the wood; botanically they are *angiosperms*, which means literally having their 'seeds in a case'. In

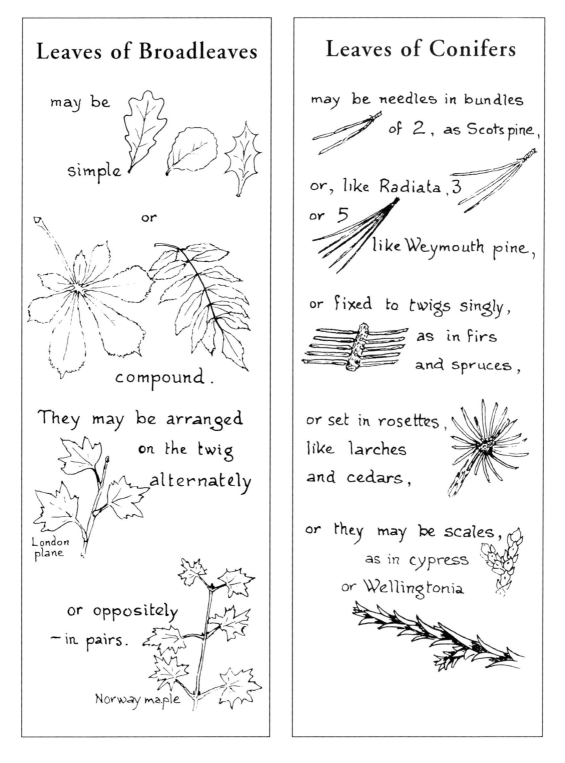

Leaves of Broadleaves

may be

simple

or

compound.

They may be arranged
on the twig
alternately

London plane

or oppositely
~ in pairs.

Norway maple

Leaves of Conifers

may be needles in bundles
of 2, as Scots pine,

or, like Radiata, 3
or 5
like Weymouth pine,

or fixed to twigs singly,
as in firs
and spruces,

or set in rosettes,
like larches
and cedars,

or they may be scales,
as in cypress
or Wellingtonia

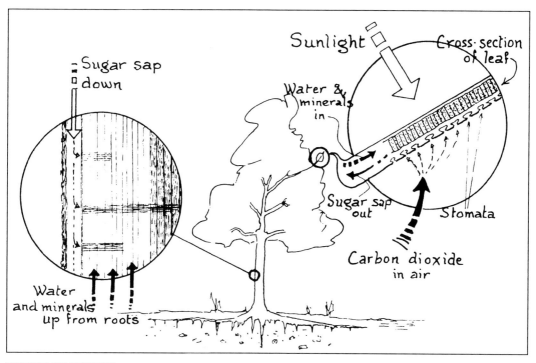

PHOTOSYNTHESIS

conifers the leaves are either needles or scales. Conifers are called *softwoods*, whether the timber is really soft or hard; botanically they are *gymnosperms*, which means 'naked seeds', the seeds mostly hanging loose in the cone.

Most broadleaves in Britain are deciduous. In autumn, as the efficiency of the leaf declines, the retrievable material in it is withdrawn into the shoot for storage, after which the leaf-stalk is sealed off and the leaf falls. In evergreen trees, which include most but not all of our conifers, each leaf remains on the tree from two to six years, after which it falls.

Whether it is after a single season or several, all trees drop a year's crop of leaves each year, which are then processed in the soil and recycled. Anything interfering with the recycling is potentially serious for the system; this includes removal of the leaves from the wood by wind, by a gardener for a compost heap or, disastrously, by tarmac over the soil surface.

LEAVES

Leaves, whether broad blades or needles, mostly comprise specialised cells containing chlorophyll, a green liquid which gives leaves their colour. When exposed to light, the cells with chlorophyll can perform the process of *photosynthesis,* in which the carbon of the air's carbon dioxide is combined with water from the roots to form starch and sugars (carbohydrates). The plant then uses these to build itself, to grow roots, new shoots, wood, bark, flowers, fruit and seeds and to form various oils, gums and resins. The air with the required carbon dioxide enters the leaf through pores called *stomata* (look with a lens on the underside of hardwood leaves and in the waxy white stripes on many conifer needles).

Photosynthesis goes on only in reasonably

strong light and in warmth, and while the leaf is internally moist. The required intensity of light varies somewhat between species: some require full sunlight; others are more shade-tolerant, a fact which has importance in the choice of trees for planting in partial shade and in managing woodland to achieve natural regeneration.

Active photosynthesis depends on large amounts of the watery root solutions coming to the leaf. The stomata, which are open to allow in the required carbon dioxide also allow much water vapour to be discharged in *transpiration*, and the air which escapes, now with depleted carbon dioxide, is oxygen-rich.

Each hectare of forest may fix about two tonnes of carbon from the air each year. (The range among common species is from one tonne per ha per annum for birch woodland to four tonnes for poplar coppice.) The total surface area of the needles and the small twigs

of conifers forms a highly efficient air-pollution filter which may adsorb each year more than 20kg of sulphur and 20kg of nitrogen (in the form of the SO_2 and the NO_x which power stations and motor vehicles discharge). The trees themselves do not pollute the environment; they clean up mankind's mess.

WOODY GROWTH

When tree shoots first grow from the extending bud in the spring they are rather fragile and 'sappy'. Then, as their growth in length is completed, the process known as 'secondary thickening' occurs. Some of the carbohydrate manufactured in photosynthesis is used selectively to strengthen the cell walls of the xylem, fixing the size of the cells and making them, for their weight, immensely strong.

Between the xylem, which carries solutions up from the roots, and the phloem,

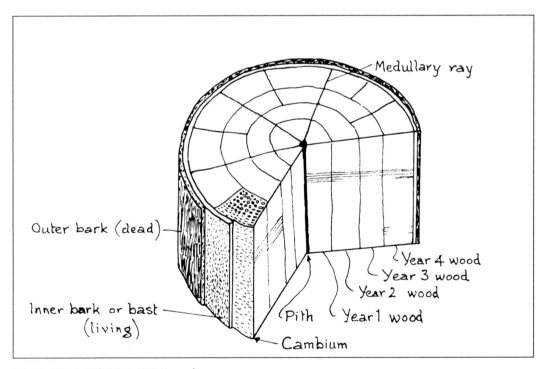

FOUR-YEAR WOODY STEM – split open

WOOD STRUCTURE

It is the secondary thickening of the wood cell walls which gives the tree its strength. The pattern of the vessels, whether the wood is ring-porous or diffuse-porous, the distribution and width of the medullary rays etc. are distinct for each species, allowing identification and determining how the wood may be worked and used.

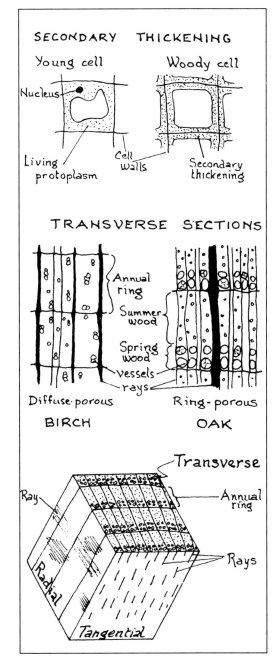

which takes carbohydrates down from the leaves, is a thin layer of actively dividing cells called the *cambium*. This forms a sheath around every twig, branch, root and all parts of the bole. On its outer side, the cambium grows cells which while living become the phloem (the bast or inner bark) and when dead the outer bark. On the inner side the cambium produces xylem cells which form the wood and also, within the wood, the specialised cells of the medullary rays (e.g. the silver grain of oak timber) which are the main stores for carbohydrates in the wood.

In the bast, and persisting in the bark, are pores called *lenticels* through which air may diffuse into the stem. The pale horizontal lines on the bark of gean and other cherries and the dark horizontal lines on silver birch bark are well-known examples.

The trees' trick of strengthening the cell walls with secondary thickening is one of the essential keys to their success: the wood allows the tree to support the crown of branches and leaves high in the air. In strength-to-weight ratio wood can scarcely be beaten, even by modern technology's new materials. Secondary thickening fixes the length of a shoot; after thickening has occurred the shoot (and the root and bole) can increase in thickness by the cambium laying down more cells on the outside of the existing wood, but it cannot increase further in length other than by growing out from a new bud at its end. Trees do not grow at all by extension of existing shoots which are woody; they do not

lengthen like squeezing toothpaste out of a tube. A branch or mark on the bole does not move up over time; it always remains at that same height.

In temperate climates the temperature

normally falls below 5°C during the winter, which causes the growth functions of the tree to close down and, in particular, the cambium cells to stop dividing. As a result, the cambial growth is intermittent, with a marked pause each cold season, and the wood it creates is in annual rings. In general, the number of rings on a cut-over stump is the age of the tree in years. The count, however, may not be exact; growth might be stopped mid-season by a cold snap or by drought, or there may be an Indian summer which restarts growth after the normal autumn, so occasionally a false ring is formed. The winter cessation of growth affects both evergreens and deciduous trees.

In spring the woody cells formed by the cambium are relatively large with thin cell walls. In summer and autumn the ratio of cell size to the thickness of the walls is smaller, so that the wood is stronger and denser. The proportion of spring wood to summer wood varies among species and from year to year. That relationship has a marked effect on the rate of volume growth, on the strength of the timber and on how it may be used and finished.

On young trees and in the stem within the live crown, the secondary thickening of the wood cell walls is thinner than in later years. This is known as *juvenile wood*. Its reduced strength is a serious matter when the timber is being marketed.

When any part of the living tree is wounded, for example by a branch being cut off, the cambium is stimulated into faster than normal growth, producing a *callus* which gradually covers the wound and seals it off. Thereafter the sheath of cambium is again continuous and the former wound is buried under fresh layers of wood and bark. If, however, the inner bark and cambium are cut through right round the stem, the flow of carbohydrates from the leaves to the roots is broken and the capacity of the roots to

supply water and nutrients to the leaves fails. The tree then dies. This wounding is known as ring-barking and may occur accidentally by animals gnawing or deliberately by vandalism.

The structure of wood varies between conifers and broadleaved trees. In conifers the most common cells are narrow cylinders with pointed ends called *tracheids*, thin-walled when formed in spring, thicker-walled in summer. The other main tissues are the *medullary rays* and *resin canals*. The medullary rays are thin vertical plates, one cell or a few cells wide, which run radially within the wood. They serve to store carbohydrates and, being thin-walled, are also important structurally by allowing elasticity in the stem and by taking up minor distortions of size due to temperature and moisture. The resin canals are fine tubes through the wood; they store and transmit resin, a by-product of the tree's growth, which seals wounds and acts as a protection against insect attacks and fungi.

Similar structures may be seen in broadleaved trees, although there are no resin canals. The main difference lies in the conduction of water; in broadleaves this is done principally in *vessels*, chains of cylindrical cells in which the top and bottom walls break away instead of thickening, so that the vessels are continuous wide vertical tubes. In some species these are wide enough to be clearly seen as pores on a smoothly cut cross-section of the wood. In some, for instance oak, the large vessels in the spring wood form a distinct line; they are known as *ring-porous* woods. Others, for instance beech, are *diffuse-porous*, the vessels being distributed across spring and summer wood.

After some years, commonly ten to twenty, the rings of wood are no longer required for the transmission of water and nutrients, their function having been taken over by later growth. The older wood then changes

BRANCH FORMATION

Branches arise from buds, so they grow from the centre of the stem, the part in the trunk being a knot. The exceptions are epicormic branches which grow from the cambium of the stem.

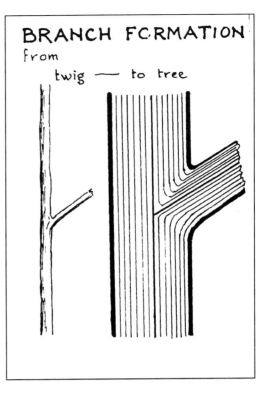

character; the tracheids of the softwoods and the vessels of hardwoods become blocked by deposits and the wood becomes darker and more durable. It is then known as *heartwood*. As timber the heartwood is more valuable than the *sapwood*, but it is not necessary to the life of the tree. It serves to give strength to the bole, but even that may not be essential, as may be seen where trees continue to stand and grow long after their heartwood, or much of it, has rotted away.

The cambium around the stem is continuous with that around every side branch and twig. As the cambium grows, the base of each branch is buried in the new wood and persists inside the stem as a *knot*. The knots derived from living branches are firmly 'grown-in' as *live-knots*. Dead branches and dead twigs, as long as they persist, are gradually included in the growing stem as *dead-knots* which are not firmly attached. Live knots, although they make the wood more difficult to work and reduce its strength somewhat, are seldom regarded as a serious defect. The presence of dead knots, however, may be a serious defect of wood, especially if it is to be sawn into planks, since they may drop out, leaving holes in each piece. More serious still are the defects arising from the enclosure of rotting branches, known as *powder-knots*.

In spite of the differences in their detailed structure, conifers and broadleaved trees function in very similar ways, both using sunlight energy to transform water and carbon dioxide into sugar or starch energy which they use to grow and flower and seed. Nevertheless, there are profound differences in the growth potential and performance of various species, where they will flourish and how they will grow.

FURTHER READING

Coutts, M.P. & Grace, J. (1995), *Wind and Trees*, Cambridge University Press, Cambridge

Hokker, H.W. (1979), *Introduction to Forest Biology*, Wiley, New York

Kimmins, J.P. (1987), *Forest Ecology*, Macmillan, New York

Pyatt, D.G. (1995), *An Ecological Site Classification for Forestry in Great Britain*, Research Information Note 260, Forestry Commission, Edinburgh

Quine, C.P. & White, I.M.S. (1993), *Revised Windiness Scores for the Windthrow Hazard Classification: The Revised Scoring Method*, Research Information Note 230, Forestry Commission, Edinburgh

Spurr, S.H. & Barnes, B.V. (1980), *Forest Ecology* (3rd ed.), Krieger, Malabar Fl, USA

Thompson, D.A. & Matthews, R.W. (1989), *The Storage of Carbon in Trees and Timber*, Research Information Note 160, Forestry Commission, Edinburgh

What Tree Is That? 3

What trees do we have?

Britain and Ireland have a small number of native trees, only about 33 in all, the count depending on what is called a tree and what is called a bush. These are the species which spread here without the help of Man after the warming at the end of the last ice age, some 10,000 years ago. Our range of trees is smaller than that elsewhere in Europe as a result of the flooding of the English Channel about 6,000 years ago; the slow travellers arrived too late to get over from France. Just how restricted the variety of our trees really is becomes apparent when it is compared with some other parts of the world; in tropical rain forest there may be 200 or more different species of trees in a single square mile, and the local people have special uses for almost all of them.

We are especially short of conifers, having only three – and two of those might be classed only as bushes. They are:

Scots pine	*Pinus sylvestris*
Juniper	*Juniperus communis*
Yew	*Taxus baccata*

The native broadleaved trees are usually accepted as:

Sessile oak	*Quercus petraea*
Pedunculate oak	*Quercus robur*
Beech	*Fagus sylvatica*
Hornbeam	*Carpinus betulus*
Silver birch	*Betula pendula*
Downy birch	*Betula pubescens*
Hazel	*Corylus avellana*

Wild apple	*Malus sylvestris*
Common alder	*Alnus glutinosa*
Field maple	*Acer campestre*
Ash	*Fraxinus excelsior*
Small-leaved Lime	*Tilia cordata*
Large-leaved Lime	*Tilia platyphyllos*
Gean	*Prunus avium*
Bird cherry	*Prunus padus*
Rowan	*Sorbus aucuparia*
Whitebeam	*Sorbus aria*
Wild service tree	*Sorbus torminalis*
Hawthorn	*Crataegus monogyna*
Midland thorn	*Crataegus laevigata*
Holly	*Ilex aquifolium*
Wych elm	*Ulmus glabra*
Goat willow	*Salix capraea*
White willow	*Salix alba*
Crack willow	*Salix fragilis*
Bay willow	*Salix pentandra*
Aspen	*Populus tremula*
Black poplar	*Populus nigra var. betulifolia*
Box	*Buxus sempervirens*
Strawberry tree	*Arbutus unedo*

This list gives both the common or vernacular name of each tree and its Latin name. You can get on very well without the Latin names but you should not be turned off by them, because they are useful and interesting. They are useful because they are international and each defines that tree exactly, without any ambiguity which might arise from the use of a common name covering several species, and many are acutely descriptive.

29

The Latin names are part of the full classification of plants which groups related plants together. For instance, all the poplars, aspen and willows belong to the Salix family, the Salicaceae, and when one looks at their features – their leaves, flowers, seeds – it is very clear they are closely related. The Latin name of each tree is in two parts, first the name of the genus it belongs to and secondly its own specific name, like a person's family name and his given name.

The botanist who has first described the plant in a scientific journal has his or her name or initials added as the author and authority. This is necessary because in many instances a plant has been given different names by different botanists collecting in different districts and sometimes a single name has been given erroneously to two quite different plants.

Quercus	*robur*	L.
Genus – *there are more than 200 oak species*	Specific name *the Latin word for oak timber*	Author citation – *this is Linnaeus, a Swede who worked in the eighteenth century, so well-known that only his initial is enough to identify him.*

Many specific names are helpfully descriptive: *Betula pendula* – the birch with pendulous twigs; *Populus tremula* – the poplar (aspen) with the trembling leaves; *Carpinus betulus* – the hornbeam with leaves like a birch.

The fact that the list of 33 trees is called 'native' and that they all reached the British Isles without help from mankind does not mean they are native to every part of Britain. For instance, beech reached the south of England on its own but was apparently introduced artificially further north and the strawberry tree is native in the south-west of Ireland but not elsewhere.

The list omits some familiar trees which, although presumed to have received help in reaching Britain, have been here a long time and are well able to sustain themselves naturally – for instance, sycamore, horse chestnut and sweet chestnut. In the conservation of ancient woodland the native trees have a special place but it would be unhelpful to restrict other planting simply to the native list on a doctrinaire basis, like woodland ethnic cleansing. The natives include only one important coniferous timber tree, Scots pine, and only a few of the broadleaves provide timber of high value. So, for producing wood to use, whether in small quantities or great forests, there is a real need to use a good proportion of introduced trees, just as there is in farm crops. For landscape reasons, too, the case for careful use of introduced species is strong; our parklands would be much less attractive if the non-native trees were absent and successful planting for shelter depends heavily upon them.

One non-native, the English elm, has been almost wiped out, apart from young suckers in hedgerows, by Dutch elm disease, a fungus carried by insects, which affects mature trees.

WHAT TREE IS THAT?

Scientists produce tabular keys – lists of critical questions about particular features – which help the reader to name an unknown plant. This is a logical way to identify specimens although it can be frustrating, especially as many plant keys depend critically on flowers:

Q: Does it have five petals or six?

A: It is now winter and I have no flowers, not even leaves, only bare twigs.

The other way of recognising trees is to use spot characters which is, after all, how we recognise our friends. In the following pages the major trees likely to be found in Britain or to be considered for planting are listed with drawings of the spot characters which are most useful in identifying them. When there are leaves on the tree these are the obvious feature to start with. In autumn the fruits are most useful but in winter the twigs and buds may be all you have of the deciduous broadleaves. When picking up fallen leaves and fruits out of season, be careful in case they belong to a tree other than the one under which they are lying.

The shape and size of the whole tree are also important clues and, with practice, even the winter colour of the twigs on leafless woodland may allow you to name the trees at half a mile or more; look for the purple colour of alder, the crimson of birch and the colour differences among the pines.

The colour of the stem bark may present difficulty. It may be quite obscured by algae, so that all the stems are green, or the stems of old trees may be covered with lichens.

The description of each tree species draws attention to the key spot features, as well as describing the natural range of the tree and the main requirements for growing it successfully – its silvicultural character. The trees do not appear here as botanists would place them, but are grouped for the convenience of the non-specialist. For instance, the London plane is grouped with the sycamore and maples because of their similar leaf shapes, even though they are unrelated botanically.

CONIFERS

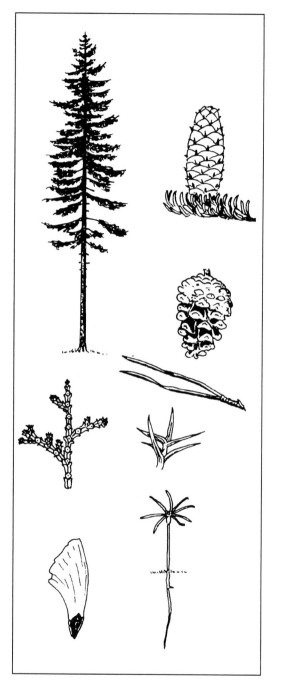

Most conifers are in the north temperate zone, forming immense forests, as in Canada and Siberia. Resinous in foliage, buds, bark and wood.

A tall, symmetrical habit and a single undivided main stem are typical. Branches generally in whorls, one whorl each year. Mostly evergreen (common exceptions are the larches). Leaves are needles or scale-like.

Separate male and female flowers on the same tree (not in yew and juniper); wind-pollinated.

Fruits are woody cones with winged seeds (not yew and juniper, which are like berries). Seeds ripen in nine to 18 months from pollination. The number of cotyledons (seed leaves) varies from two to 15 or more. Most are difficult to grow from cuttings. Sprouts and root suckers are very rare.

Timbers ('softwoods' in the trade, although some are hard) used for construction, fences, boxes, particle-board, hardboard, poles, pulp for paper. World demand for industrial woods is mainly for conifers (long straight timbers, excellent strength-to-weight ratio and good for paper-pulp). The imposing form and evergreen stateliness make many conifers valuable for landscape planting.

Top right: two types of cone
Centre: three types of foliage
Bottom: a typical winged seed and
a seedling with cotyledons (seed leaves)

Taxus baccata YEW

Native to central and southern Europe and all Britain except north Scotland. A low-growing evergreen tree up to 20m, densely branched. Slow-growing and may produce many basal shoots to form a massive fluted trunk. The bark is rust-red, peeling in long strips.

The needle-leaves are set spirally on the shoots but, each with a basal twist, they form two ranks on all horizontal branches. Each needle may be 1–3cm long, ending in a horny point; glossy bright green above, pale green below. The buds are small, fat and green, with leafy scales.

Male and female flowers grow on separate trees. The fruit is a pink fleshy cup (pea-size), almost enclosing a greenish bone-hard seed. The flesh is soft, sweet and non-toxic, very attractive to birds which distribute the seed. The bark, wood, needles and seed contain the poison Taxin. Half-withered foliage is particularly poisonous to horses and cattle.

Yew's shade-tolerance exceeds all other forest trees. It grows best on a limy loam soil and in sheltered places. Very long-lived.

A valuable and decorative timber, tough and resilient; deep red-brown heartwood and contrasting white sapwood; used for cabinet-making and turnery, formerly for long-bows. An excellent hedging plant with a long history of planting in churchyards. There are seven Taxus species and many garden varieties.

SILVER FIRS

Abies species

There are more than 40 Abies species, all found in the north temperate zone. They are evergreen conifers, typically with tall undivided main stems and crowns which flatten at the top in later life. Branches are in well-defined whorls, one each year. All are very resinous.

Each needle is set on the twig with a 'sucker' base which leaves a smooth circular scar when the needle falls at five years or so (cf. the woody pegs of spruces). The needles, rarely spine-tipped, are set spirally on the twig but in most species are twisted to form two flat combs, but sometimes hiding the top side of the shoot. Buds are short, fat and mostly resinous, usually hidden by the needles.

In all Abies the cones are erect on the twigs, some species very large and some with projecting papery scales (bracts) interleaved with the cone scales. When ripe, the cones break up on the tree, releasing the winged seeds and leaving the central stem as an upright spike (may be the origin of candles on Christmas trees).

European silver fir (*Abies alba*) is now rarely planted in the UK, tending to suffer severely from aphid attacks, late frost and fungus diseases. Some North American species grow well. Caucasian fir (*A. nordmanniana*) is important as the top-market Christmas tree and for greenery. Silver fir timber is generally less dense, rougher and less valuable than pines and spruces, used for rougher joinery work, paper-pulp, particle-board, etc.

Abies grandis

GRAND FIR

Western North America from Vancouver Island to California and inland to Montana and Idaho. A tall, handsome evergreen, reaching up to 90m in its home and 60m in Britain. Introduced in 1831 by David Douglas.

The needles are bright glossy green above and have two white bands beneath. They are up to 5cm long, clearly notched at the tip; those growing from the top of the shoot are shorter than the others, but all are twisted at the base into two flat ranks, leaving the top of the shoot clearly visible from above. Each needle's sucker foot, when the needle falls, leaves a smooth circular scar. The crushed needles emit a delicious resinous aroma, like tangerine oranges.

The young bark is smooth with many resin blisters; it ages to deep brown and fissured, later grey. Cones grow erect to about 10cm tall, breaking up whenever the seed is ripe.

The tree is shade-tolerant and fast-growing (once established and protected from browsing animals), thriving on moist loam soils. In sheltered favourable sites it can produce a very large volume of timber; a stand in Perthshire appears to be the fastest-growing stand in Britain, at an annual mean volume growth of 34m³ per ha.

The timber use is limited, because of the large amount of juvenile wood, to boxwood, pulp and particle-board. It is a fine specimen tree in parks.

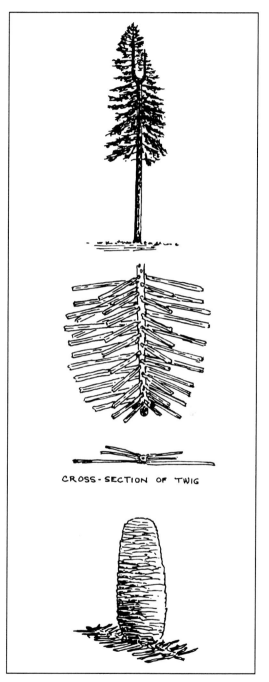

CROSS-SECTION OF TWIG

NOBLE FIR

Abies procera

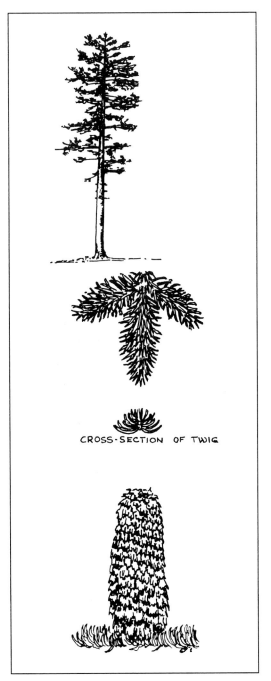

CROSS-SECTION OF TWIG

Evergreen conifer from north-west USA, in the Cascade Mountains and coastal Oregon. Introduced by David Douglas in 1830.

The tree grows upright in severe wind exposure; a main use in the UK is to form windfirm forest edges and in shelterbelts.

Viewed from above, the twig is quite hidden by the curved grey-green needles. The buds are small and round, tipped with resin, the bud scales free at their tips.

The cone is massive, 15–25cm tall, with bracts interleaved with the cone scales, giving a ragged appearance. The cones disintegrate as soon as the seed is ripe, making the seed difficult to collect. Seed is best sown immediately on collection.

The wood is good, pale brown in colour. The bark of young trees has resin blisters. The bark of older trees is pale grey.

This is a very decorative tree, excellent in parks and on woodland edges both for its appearance and its shelter effect. Silviculturally the noble fir likes deep, sandy loam soils with a good water supply, but it grows moderately well on acid mineral soils and shallow upland peats. It avoids lime. An important shelterbelt tree, even in severe exposure in the hills, and gaining popularity and importance as a Christmas tree.

Pseudotsuga menziesii

DOUGLAS FIR

Western North America, along the Pacific coast and Cascade Mountains from British Columbia to Mexico, but only seed sources from Oregon northwards are useful in Britain. One of the world's most important and magnificent timber trees, in its home growing up to 100m tall and to 800 years of age; now the UK's tallest trees at 65m.

The tree is easily recognised by its buds and cones. The buds are shining brown, scaly and taper to a fine point, similar to those of beech; this distinguishes it from other conifers. The needles form two flat ranks, deep green above with a marked groove; below are two grey bands (not white) either side of the midrib.

The pendulous cones are 6–7cm long and pale brown. Outside each cone scale there is a three-pointed bract, a unique feature.

The young bark is smooth olive-green with many resin blisters; later it is dark with deep vertical fissures, often showing the golden inner bark at the bottom of the cracks.

Douglas fir tolerates a wide range of rainfall but is discriminating in soil. Unsuited to heathland and moorland. Deep-rooting, it thrives on well-drained mineral soils but dislikes heavy clay and needs at least modest fertility. It does not stand severe wind exposure (the crown deforming and tattering). It tolerates side shade in youth. Very productive on suitable sites. A tree for replanting, not a pioneer.

The timber is one of the finest in the world, used for all kinds of construction, flooring, furniture and plywood, both decorative and construction grades.

Named after David Douglas, the Scottish plantsman who first sent seed home in 1827, and Archibald Menzies, the Scottish botanist who discovered the tree in 1791.

WESTERN HEMLOCK *Tsuga heterophylla*

Western North America from Alaska to California, along the coast and on the lower slopes of the Cascade Mountains.

This is a graceful evergreen, with a narrow crown when mature, growing as a natural associate of Douglas fir, western red cedar and Sitka spruce.

The tree is easily recognised by its drooping leading shoot and by its foliage. *Heterophylla* means 'with different leaves' and the needles are distinctly of different lengths, those on top of the shoot half the length of the others, 7–12mm mixed together. The young shoots are slender and drooping, yellow-brown and hairy. The small round buds are brown and hairy.

The small egg-shaped cones are very numerous, 20–25mm long, and the cone scales longer than broad.

Western hemlock accepts a wide range of rainfall but thrives best on mineral soils in moist climates in western Britain. It is shade-tolerant and will grow on wet clay soils but is not suited to exposed sites and frost hollows. It produces abundant seed and can be very invasive, regenerating freely, although it is prone to Fomes stem-rot, which severely limits its value. The white timber is useful for pulp wood and particle-board. In North America it is used as a general purpose saw timber, similar to spruce. It is very prone to deer-browsing.

Picea species

SPRUCES

There are about 40 species in the spruce genus, all in the temperate and colder regions of the northern hemisphere. They are erect evergreen conifers with undivided straight stems and pointed crowns.

All spruces have the needles each standing on a tiny wooden peg, set spirally on the twigs. When the needles fall (after four or five years), the pegs remain, leaving the twigs very rough (cf. the smooth twigs of Abies).

The cones are pendulous, the cone scales opening on the tree to release the winged seeds and then usually persisting on the tree for some time and falling entire. Some species have cones which are leathery in texture; others have papery scales.

The timbers of all the spruces are similar: light in colour, smooth-textured and with an excellent strength-to-weight ratio, used for building construction and all kinds of joinery. Several are long-fibred and, for that reason, valuable in paper-making to give high tear-strength.

The pointed crown of the spruces is a marked feature. In parts of Germany annexed by Prussia, where the Norway spruce is not native, it is known as *Preussenbaum*, 'the Prussian tree' – jocularly after the spike on top of a Prussian officer's Pickelhaube.

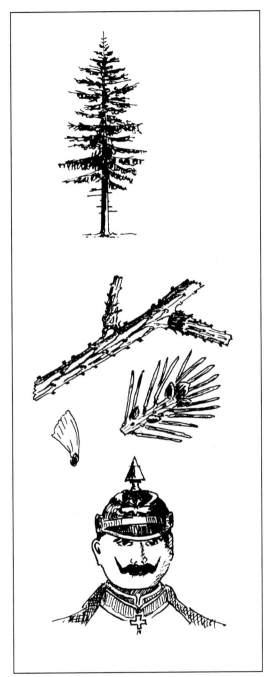

NORWAY SPRUCE

Picea abies

Central Europe, Scandinavia and north Asia. Introduced to Britain in the sixteenth century. An evergreen with a tall pointed crown. Whorled branches but with odd branches between.

The bark is smooth orangey-grey, flaking off in small, thin scales.

The needles are fresh mid-green colour, 2cm long, blunt-pointed and set on tiny wooden pegs. Since the needles are blunt-pointed and point 'forwards', the twig can be grasped in the bare hand without pain (cf. Sitka spruce).

The cones, 8–11cm long, hang downwards and do not break up on the tree (cf. Silver firs). Seed is fertile and freely produced, but the cone scales are leathery and may not open properly in our mild, damp winters. Very winter-hardy and less affected by late frosts than Sitka spruce. It thrives on high humidity and requires at least modest fertility; grows best on moist loams but tolerates flushed peat if surface drainage is good. On heathery sites it is liable to 'check' severely. Naturally shallow-rooted, it is susceptible to windthrow, drought and soil compaction. Sensitive to air pollution and salt. Tolerates partial shade in youth.

Excellent timber, mild-textured and strong for its weight; imported as 'Baltic whitewood' or 'white deal' from Scandinavia for construction and carpentry. The long fibres make it valuable for paper-pulp. It has been Britain's traditional Christmas tree.

Picea sitchensis

SITKA SPRUCE

A narrow coastal strip of North America from California to Alaska. The largest of the spruces, this magnificent tree was introduced to Britain by David Douglas in 1831. For 70 years it was a park tree only, but since then it has been a major plantation tree.

The bark is duller than Norway spruce's and is shed from older stems in large flakes.

The needles are stiff, 2cm long and sharp-pointed, bluish-green above and, usually, marked white bands below. Set at about right angles to the shoot, the sharp needles make it painful to grasp the shoot in the bare hand. The foliage is not favoured by browsing animals.

The cone is shorter than Norway spruce's, 6–8cm, with dry papery scales (not leathery like Norway's). The cone opens readily to release the seed. On suitable ground it regenerates freely.

It is not necessarily shallow-rooted but it tolerates shallow soils and can grow on drained peat sites. It requires moisture and should not be planted in less than 1,000mm rainfall (40 inches). It stands severe wind exposure, including salt, and is fully winter-hardy but sensitive to late frosts. It is intolerant of shade. Its capacity to grow superbly on fertile sheltered sites and reasonably even on infertile, exposed ones puts this tree in the first rank for forestry plantations. It produces excellent timber, used for building construction, pallets, paper-pulp, particle-board and, in its home range, highest-quality plywood.

Its high yields of valuable timber give Sitka spruce great importance in UK and Irish forestry. Gives quick shelter. Unsuitable as a Christmas tree because of the risk of eye injuries by the sharp needles.

LARCHES

Larix species

The genus comprises ten species in cool and mountainous parts of the northern hemisphere, none native to Britain.

Larches differ from other common conifers in being deciduous, although first-year seedlings may retain their needles over winter. The needles on twigs not more than one year old are always solitary, set in an open spiral round the shoot. On all older shoots, the needles grow in tight bunches on little knobs, short shoots which never grow longer.

The larches are also unusual in having showy female flowers, like pink roses, beautiful among the fresh green needles in spring. The cones are leathery, erect on the twigs, persistent for several years and difficult to open to release the seed. Crossbills may rip open the cones and thereby help natural regeneration.

Larch timber is denser than many other conifers and more durable, although difficult to season without splitting. Naturally durable, it is used for fencing timbers, posts and, traditionally, boat building. As firewood larch is dangerously sparky.

The sketches are of European larch, the upper a parent larch at Dunkeld Cathedral, planted in 1738 at the main introduction of the species to Scotland when the 4th Duke of Atholl used it for a 4,000ha afforestation project.

Larix decidua

EUROPEAN LARCH

Central Europe, the Alps and Carpathians. A tree of the mountains and high hills, usually with a clear bole and graceful downswept branches. First introduced to Britain after 1600 but planted on a large scale only in the mid-1700s.

Bark on old trees is deeply fissured, red-purple, often encrusted with lichens.

The twigs are straw-yellow in colour and the slender needles a clear fresh green (not bluish), 2.5cm long. The upright cones are tall and oval, with straight cone scales (not turned over at the tips). These three features distinguish this species from Japanese larch and hybrids. In autumn the needles turn golden. The cones are about 3.5cm tall by 2cm in diameter.

Larch is very light-demanding and grows quickly in youth, requiring ample space and light; thinning should keep a live crown at least 40 per cent of its height. It needs freely draining light mineral soils and thrives in a dry Continental climate (more so than Japanese larch). Trees are liable to attack by a wound parasite fungus causing stem cankers and to crown die-back. The seed sources most suited to Britain are Poland and Sudeten foothills. Seed from the high Alps is quite unsuited to the maritime climate, trees suffering severely from diseases.

The sapwood is pale, the heartwood distinctly reddish-brown and durable, making it especially valuable for fencing and other outdoor work and for boat-building.

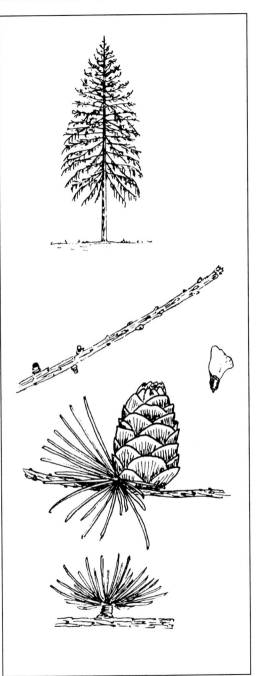

JAPANESE LARCH

Larix kaempferi

Native to Japan, the central mountain area of Honshu Island, around 2,000m, on rich volcanic soils. It was introduced in 1861.

In youth it grows more quickly than the European larch and does not suffer from stem canker. The habit differs from European larch; branches grow horizontally and spread wider; it casts heavier shade.

The twigs are glistening red and the needles bluish-green (glaucous), 3cm long and slightly stouter than those of European larch. The upright cones are spherical, with the tips of the cone scales sharply bent back. These features clearly distinguish Japanese from European larch. In autumn the needles turn pale brown rather than the gold of European.

The rapid growth of this tree and its resistance to the canker disease made Japanese larch seem preferable to European but unfortunately the timber quality is poorer, in both the shape of the stem and loss of durability. Its cultural needs are similar to European but it requires moister and milder conditions; it tolerates heathland conditions and slightly more shade.

On fertile soils, the stem of Japanese larch has a tendency to grow spirally rather than straight, a serious fault for timber production. Much was planted from 1930 to 1960 on steep brown earths in uplands of Wales and west Scotland.

Larix x eurolepis

HYBRID LARCH

In 1897 plants raised from seed collected from the Japanese larch avenue of Dunkeld House in Perthshire were planted nearby and later seen to be unlike true Japanese larch seedlings. The parent female flowers had been fertilised by pollen from the magnificent European larch nearby. The young plants were hybrids.

In form (straight bole and sweeping branches) and timber quality, the hybrid closely resembles European larch. The rate of growth and general vigour exceed even Japanese larch and the clinching benefit is freedom from larch canker and die-back. The hybrid inherits the good qualities of both parents, and hybrid vigour in addition.

The twigs are light-copper-coloured. The needles are longer and stronger than European, but richer green. The upright cones are taller than broad but have tips slightly bent back. In all these characters, the hybrids are intermediate between the parent species. Nowadays the crossing is not left to chance but is made in seed orchards planted with selected 'plus-tree' parents.

In forest planting hybrid larch has replaced both its parents. It is intolerant of shade, requiring plenty of crown space; it thrives on well-drained mineral soils. It is a useful nurse species for some broadleaved trees and in landscape design, its fresh green foliage brightening areas of darker green conifers.

CEDARS AND DEODAR

Cedrus spp.

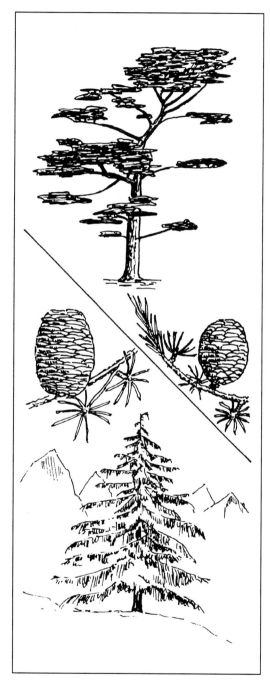

Three Cedars are grown in Britain:
Atlas cedar *C. atlantica*: Atlas Mountains
Cedar of Lebanon *C. libani*: Asia Minor
Deodar *C. deodara*: North-west Himalaya

Evergreen conifers, all very similar. Conical form when young, all later develop massive boles and tabular crowns. The bark is smooth grey when young, becoming scaly and eventually deeply furrowed.

The foliage closely resembles larch but is evergreen. On the long shoots the needles are individual and spiral, but most are in rosettes of 20 to 40 on short shoots. Many cultivated varieties, especially with blue-green foliage. The cones, which take two years to mature, are massive, barrel-shaped and upright on the branches. When the seed is mature the cones break up on the tree.

In their homes all three species are important timber trees, for construction, joinery and railway sleepers. In Britain they are parkland and landscape trees.

Cedar of Lebanon needs mild sheltered sites with moist, fertile soil; it tolerates lime but is frost-tender. Needles 3cm; cones about 10cm. Leading shoot not pendulous (top diagram).

The deodar has longer needles than the others (up to 5cm) and a larger cone (up to 13cm). Leader pendulous. It is best on lime-deficient, moist, well-drained soils but is susceptible to unseasonable frosts (lower diagram).

An imperfect but generally good guide:
Atlas twigs are **Ascending**
Lebanon twigs are **Level**
Deodar twigs are **Drooping**

Cedrus atlantica

ATLAS CEDAR

North Africa, from the Atlas Mountains above 1,000m elevation.

A handsome tree with branches set at a sharp angle to the stem.

In the wild there is variation in foliage; the commonest form planted in Britain is the variety 'glauca', which has grey leaves. The needles are 2–3cm long and the cones 7–9cm (both smaller than Lebanon and Deodar). Typically branches are ascending at the tips (*mnemonic*: Atlantica – Ascending).

More commonly planted and altogether more robust than the other cedars. It is frost-hardy, not sensitive to summer droughts and relatively insensitive to smoke pollution. It is said to prefer limy soils but will grow on acid brown earths. It is useful in mixture with beech on dry grass sites in low-rainfall areas.

The Atlas cedar, like the other species of Cedrus, at home is an important timber tree but in Britain is grown mainly as a park specimen, valued for both its crown shape and its grey colour. A strikingly handsome tree.

'Cedar wood' of commerce is usually not from cedars but from Cedrela, a tropical broadleaved genus. *Cedrela odorata* is the cigar-box cedar, *C. mexicana* the Honduras cedar. Yellow cedar is marketed from the Nootka cypress *Chamaecyparis nootkatensis*, a close relation of Lawson cypress, and white cedar from a thuja, both conifers.

PINES

Pinus species

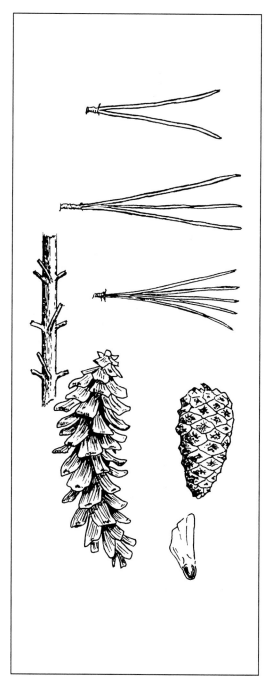

There are some 80 pine species, all in the northern hemisphere, from the Arctic to the subtropics. Only Scots pine is native to Britain. They are evergreens with pointed crowns in youth, tending to flatten later. Side branches grow in neat whorls, usually with no intermediate twigs. Most are very resinous, some regularly tapped for resin and turpentine. They are very important timber trees worldwide, for building construction, joinery, poles, paper-making, etc.

All pines have needles set in small bundles, pairs, threes or fives (really short shoots with a bud between the needles). The important timber pines in the UK all have needles in pairs and hard woody cones which open on the tree to release the winged seed.

As well as the three on the following pages, *Pinus pinaster* is a Mediterranean two-needled pine, widely planted on England's south coast, and in south-west Ireland. Much used for fixing sand-dunes and is tapped for resin and turpentine (e.g. the Landes in south-west France). Up to 30m tall; needles up to 20cm, dark green. Cones are the largest European pines, 15cm, shining bright brown.

Occurring only as specimens and park trees in Britain but immensely important for timber elsewhere in the world:

Pinus radiata (three needles): native only on the tiny Monterey peninsula in California, forming huge plantations in New Zealand, Australia, South Africa and Chile.

Pinus strobus (five needles): Weymouth pine (white pine) from the eastern USA.

Pinus ponderosa (three needles): Western yellow pine in western USA, growing to 70m.

Pinus sylvestris

SCOTS PINE

Native to UK (Scotland only), Scandinavia, Central and Eastern Europe and Asia.

Tall, normally undivided main stem; crown pointed when young, the top flattening with age. The only common tree with a distinctly orange bark, especially on the upper bole, shed in papery scales. At the base old trees have thick pinkish-grey bark, deeply fissured into broad flat slabs.

Needles are in pairs, blue-green and short, about 4–6cm. Buds are narrow, blunt-pointed, usually resinous, with the bud scales free at the tips.

Mature cones are first green and tightly closed, later brown when they open to release the winged seeds. Cones are hard and woody, 4–6cm long, the scales having no spine at the end.

Scots pine is a marked light-demander and a pioneer, at its best in the sunnier and drier north-east rather than in the west. Not sensitive to frost or heat. Modest in its soil requirements; best on freely drained gravels and glacial sands, but hardy and accommodating. On shallow calcareous soils it has a short life. Often used as a nurse species, especially for oak on light soils.

The excellent timber has a reddish, resinous heartwood, not naturally durable. It is imported as 'redwood' or 'red deal' and extensively used for construction, joinery, poles, flooring, etc., and also for particle-board and paper-pulp. A major commercial Christmas tree in the USA and gaining popularity in parts of the UK.

The main tree for reconstituting the Caledonian pine forest in Scotland, for which great attention to seed source is required.

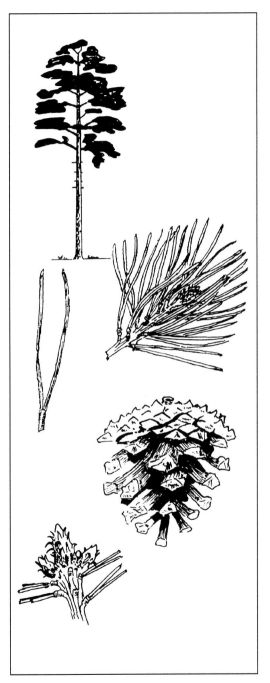

CORSICAN PINE

Pinus nigra subspecies *laricio*

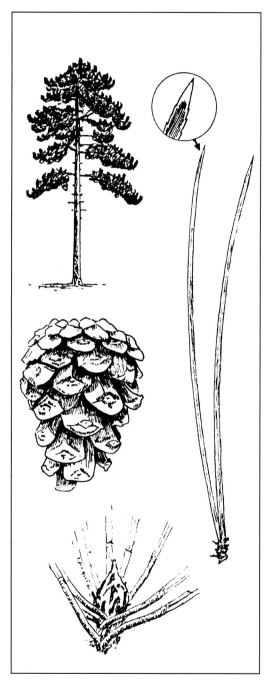

One of several local races of the black pine, *Pinus nigra*, widely distributed in southern Europe and Asia Minor; introduced to Britain in 1759.

Grows up to 35m with a straight bole and a compact columnar dark green crown. Twigs of the whole crown are usually upturned, even the lower branches.

Bark is dark grey throughout, and in old trees deeply fissured; the absence of orange colour immediately distinguishes this from Scots pine.

Needles long and flexible, deep green in colour, 10–15cm long and not more than 2mm broad, ending in a sharp, horny point. Buds are onion-shaped, 10–15mm, resinous and with scales closely pressed in.

Cones are woody, 5–8cm, like Scots but larger and usually curved.

This pine grows fast in lowland sunny areas with low summer rainfall; frost-hardy and tolerant of salt; windfirm and drought-resistant. Prefers base-rich light soils, and warmth. Light-demanding. The timber has uses similar to Scots pine but is less well regarded; heavier, more resinous and forms heartwood later.

The closely related Austrian pine is useful as it succeeds on exposed limestone and chalk sites and even on dry, shallow soils, where it is a good nurse tree for beech.

Pinus contorta

LODGEPOLE PINE

Western North America, from Alaska to Colorado, both on the coast and high into the Rocky Mountains. There are many local varieties or provenances; the Native Americans used the tall straight poles of the inland strains for their lodges or wigwams, but there are also bent-boled and wide-crowned strains from the coast.

The bark is thin, scaly and light grey; even in old age it is dull grey and forms small, squarish plates.

The needles, in pairs, are rather stiff, sharply twisted ('contorta'), strikingly bright green and minutely saw-toothed on the margins; 3–5cm long. On some races male flowering is so prolific that after they are shed there are bare patches on every twig, so obvious as to identify the tree.

The egg-shaped cones are small (6cm) and woody; each scale has a small, sharp prickle at the end (cf. Scots pine). They often grow directly on the stem and may not open even in very dry weather but in natural forest they rely on fires to dry them and release the seed.

The tree is a pioneer, winter-hardy, resists extreme exposure, including sea winds, and, with regard to soils, is very undemanding, from dry soils to poor peat. Coastal varieties, planted in mixture with Sitka spruce, suppress heather and establish the latter. Pine beauty moth may attack it (see chapter eight); it is also favourite deer food. The timber is used for particle-board and paper-pulp; it takes a fine finish but is regarded as inferior to Scots pine, especially the coastal provenances, the boles of which are usually bent and have both compression wood and large knots.

WELLINGTONIA

Sequoiadendron giganteum

Western North America in small groves on the western slopes of the Sierra Nevada in north central California. Commonly planted around 1860 as an avenue and park tree in Britain, now forming massive tapering boles, broadly buttressed and renowned for the spongy cinnamon-coloured outer bark.

The crown is a narrow spire, with graceful downswept branches. The evergreen foliage is entirely of sharp-pointed scales, the tips being free. The cones are barrel-shaped and about 5cm long, with some spiked bracts projecting when fresh.

At home these trees grow to a great age — some over 3,000 years. Part of their secret of longevity is the spongy bark which is fireproof and up to 60cm thick, insulating the cambium from heat damage. Also important are the flexible downswept branches; they both shed snow, the cause of breakage in many other species, and easily deform in storm winds, 'streamlining' the crown and preventing damage. Even in severe exposure, this tree appears to be virtually immune from windblow.

Although mature timber in the USA is excellent, the tree is grown in Britain in avenues and parks for its appearance. It is best on loams and brown earths, including acid ones. It is hardy and has been planted throughout Britain except in the northern isles; the best specimens are in the west and there is a particularly fine avenue at Benmore Gardens in Argyll.

Sequoia sempervirens

COAST REDWOOD

Western North America, in a narrow strip on the Pacific coast in southern Oregon and northern California.

An extremely tall evergreen, with a narrow spire crown, a sharply tapering bole and spongy, cinnamon-red, fireproof bark.

This is easily distinguished from the wellingtonia by the foliage. The leaves are of two types: the main twigs and the shoots carrying the cones are covered with scale-leaves clinging to the twig but free at the points; the side twigs carry flattened needles which are graded in length from 2–3mm increasing to 12mm and then tapering down again to 2–3mm. The cones are small, woody and barrel-shaped, about 10mm long.

The coast redwood grows well only in a moist oceanic climate and in Britain better in the warmer south. It is remarkable as one of the few conifers which grows easily from cuttings and which puts up root suckers and even coppice-shoots.

Although it has been grown occasionally as a plantation for timber, producing high volume, it is much more common as a specimen tree. It will grow on limy soils but its requirement is really for fertile mineral soil and a moist climate. It is frost-tender.

WESTERN RED CEDAR

Thuja plicata

Western North America, from sea level to 2,000m in the Coastal Range and Cascades; introduced to Britain in 1843. An evergreen with a straight bole which tapers sharply. It has thin cinnamon-brown bark which peels in strips.

The foliage comprises small overlapping scales pressed closely on shoots to form flat fern-like sprays, which may be confused with the cypresses, but they are easily distinguished by the cones.

The scale-leaves have oil glands and when crushed give off a powerful smell (cypress has a lighter resinous smell). The foliage is glistening bright green and the tips of the fronds are fleshy (Lawson's are typically thinner and duller bluish-green).

The cones are 1cm long, with six or eight scales in opposite pairs, each a narrow oblong in shape. They are leathery, green when immature, later mid-brown when they open to shed the double-winged seed.

The tree is shade-tolerant, requiring deep, moist, fertile soil for good development; it grows on calcareous soils but is not at home on sandy and acid soils. Its shape and foliage colour make this a useful tree for landscape design. An excellent hedge plant.

The timber is very lightweight and extremely durable, red-brown when freshly cut, weathering to silver-grey. It is used for shingles, weather-boarding and fencing. Cut foliage is used for decoration. The resin contains a powerful insecticide.

Chamaecyparis lawsoniana

LAWSON CYPRESS

Western North America, in a small coastal strip in south Oregon and north California. The genus has six species in North America and East Asia. An evergreen with a pendulous leading shoot and the bole often forking. The branches are short and pendulous, forming a narrow, blunt-topped crown. The bark is thin.

Like thuja, cypress has foliage of scale-leaves forming fern-like sprays. The upper surface is typically deep-green or bluish-green, the lower bloomed with white wax. There is a resin gland on each scale, giving the crushed foliage a sharp resinous aroma. The tips of the foliage sprays are thin (thuja's are fleshy).

The cones are quite distinct: spherical, about 8mm across, woody with spikes on the tips of the nine or so scales. They change from, briefly, creamy-green to blue-grey and finally reddish-brown in October, when they release the double-winged seeds.

Lawson cypress likes a moist, sandy loam in some shelter, and at least moderate soil fertility; it is semi-shade-tolerant. An ornamental, not for production forestry. More than 100 garden varieties, including many dwarfs. It stands clipping as a hedge plant. The foliage is readily marketed for decoration. The yellow timber is lightweight and naturally durable. The tree is named after Peter Lawson, a forest nurseryman in Edinburgh, to whom seeds were first sent in 1854.

The Leyland cypress, x *Cupressocyparis leylandii*, is a hybrid between Nootka cypress and Monterey cypress. It grows fast, its only merit. It is quite unsuited for planting in town gardens, where it quickly gets out of hand (expect it to grow to 40m (130 feet) on garden soils); an unattractive tree but tolerates hard exposure.

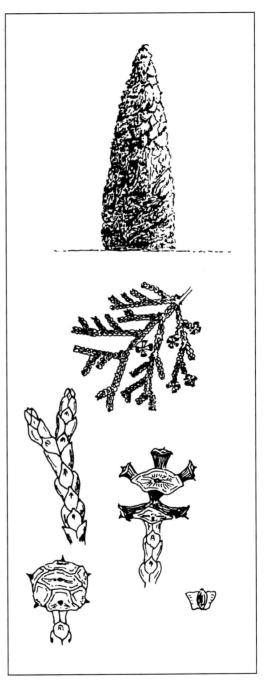

JUNIPER

Juniperus communis

Native to Britain and a wide distribution across Europe, Asia and North Africa. In Britain it occurs on some chalk downs in England and is common in relict native pine and birch woods in Scotland and relict oak woods in Cumbria and Scotland on acid gravel soils.

This is an evergreen shrub, occasionally up to 9m in height, growing slowly and living to a great age.

The needles are tapered, sharp-pointed, grouped in twos or threes and blue-green in colour. The buds are small, without scales and lost in the developing foliage.

Male and female flowers occur on separate plants. The fruit is a pea-sized berry, at first green, then dark brown to violet. It fruits annually, the seeds normally resting for a year before germinating.

Juniper is very modest in its soil requirements and will grow on glacial sands, loams and chalk, liking sunny places. It is frost-hardy.

The wood is decorative, tough and durable. Although it grows only to a small size, it is valued for inlay work and turnery. The berries are used medicinally and as a spice. They are used to flavour spirits such as gin, Dutch genever and German Steinhäger.

ANGIOSPERMS (meaning 'seeds in a box')

These trees are arranged in more families than the conifers. The key features for botanical classification are the flower structures but the important species in Britain can be readily identified in summer by the arrangement and shape of the leaves, and in winter by the buds and the form of the tree. All our broadleaves have two seed leaves (cotyledons).

Most broadleaved trees in Britain are deciduous, shedding their whole suit of leaves in the autumn and growing a new one each spring. A few are evergreen. All are called 'hardwoods', irrespective of the actual hardness of the timber. Some, like oak and beech, are indeed quite hard, but others, like poplar and willow, are soft in texture.

In contrast to the conifers, the great majority of broadleaves can be propagated easily by vegetative means as well as from seed. Most broadleaves sprout from the cut-over stem (coppice) or from the roots, although not with equal facility.

The growth habits of the two groups differ markedly. The height of the main bole of a conifer on a given site is almost the same whether it grows in a closed stand or is isolated. In contrast, virtually all broadleaves, if grown in an open position, develop only a short bole and then branch into a wide crown; they develop a tall stem, most valuable for timber, only if grown in competition with neighbours in a closed stand.

BROADLEAVES

SESSILE OAK

Quercus petraea

The most important broadleaved tree in Britain and Ireland, growing up to 30m in height on the best sites. Native to all Europe except northern Scandinavia and southern Spain.

Distinguish from pedunculate oak by its straighter bole, typically continuous from the base to the top of the tree, and more regular branching; the stalked leaves mean the foliage, when one looks up, is more evenly spread – less clumped – than that of pedunculate oak.

Leaves are stalked, the leaf-base wedge-shaped and without folded ears; the lobing is regular and the blade widest nearer the tip than the base. Veins run to the lobe points, not to the bays. The leaves are leathery, with star-shaped hairs on the lower side (see with a lens). Buds are rather slender, pale brown and pointed. The fruit, acorns, are without stalks (i.e. sessile). Acorn size varies widely.

Grows well only on deep, well-drained soils which are not too base-rich; better than any other tree, except pedunculate oak, on heavy clays, but avoids wet soils, especially if liable to flooding. It grows on acid brown earths on glacial gravels. Less frost-tender than pedunculate oak and less prone to epicormic branching. Deep-rooting and a soil-improver. Very wind-resistant. It coppices well. Formerly large areas were managed as coppice to provide bark for tanning leather and charcoal for metal-smelting.

The heavy heartwood is an excellent timber for construction, fencing, gates, furniture and, in the largest sizes, veneering. Good-quality logs fetch high prices but poor grades only one-tenth of the best.

As sessile acorns are less easy to store than pedunculate, nurserymen long favoured pedunculate oak, and sessile woods were adulterated. The better timber qualities of sessile are now recognised along with its greater suitability for most available oak sites.

Quercus robur

PEDUNCULATE OAK

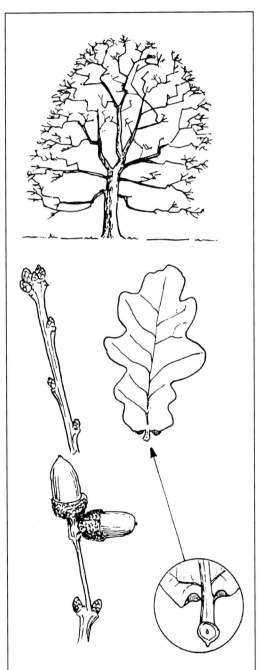

Native; the range is similar to that of sessile oak but it extends further into middle Russia.

Although similar to sessile oak, this is a less desirable and less valuable species for most sites. The branches are more contorted and the bole is seldom continuous through the crown.

The leaves have almost no stalk and at the base the blade has two 'ears' (auricles) folded over. The lobes of the leaf are often irregular and the widest part of the blade is at the middle. Veins run to the bays as well as the points of the lobes. The leaves are more papery than sessile's.

The buds are pale brown, plump, blunt-pointed and clustered at the tip of the twig. Following a tendency for a side bud to develop, the twigs are often zig-zag. The acorns have long stalks (peduncles).

Silviculturally this tree is similar to sessile oak, but it tolerates heavier clay soils and wetter conditions; it is also more frost-tender and benefits on many sites from the use of a nurse tree (alder, pine, etc.). The bole is liable to frost-cracking and, on thin, dry soils, to drought cracks.

This oak also produces excellent timber, although the tendency for the bole to grow abundant adventitious shoots in middle life reduces the quality. It coppices well and large areas were so managed prior to 1900. Over centuries there has been some genetic mixing of sessile and pedunculate oaks, so many trees show features of both, e.g. sessile acorns but leaves with auricles.

TURKEY OAK

Quercus cerris

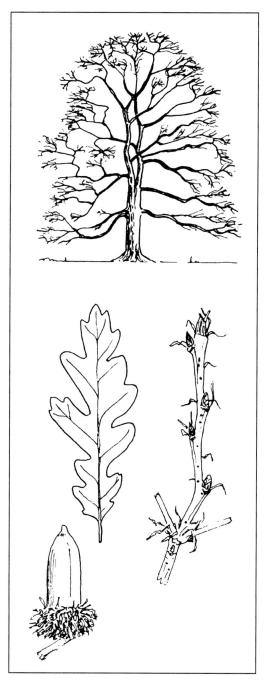

Southern Europe, Asia Minor and the Mediterranean islands; introduced to Britain in about 1730 and widely planted in parks in the nineteenth century.

A tall, handsome tree, growing to over 30m in south-west England and almost that size in Scotland as far north as Inverness.

The leaves vary in size and shape from 7–15cm in length, generally narrow, 1–4cm, with irregular sharp-pointed lobes and often very 'narrow-waisted'. They are dark green above, downy and pale below. Buds have long, thread-like scales, and young twigs carry these also.

The tall, sharp-pointed acorns are held in shallow cups which are covered with narrow, curly scales looking like hairs. The acorns take two years to ripen.

The bark is similar to the native oaks, although darker. The timber is heavy and has a poor reputation, although whether this is justified is unclear. It certainly tends to have a wider soft sapwood than our native oaks and the heartwood may be less durable, although it takes a good finish.

A park and shelterbelt tree, valued for its ability to grow on heavy, lime-rich soils and for its windfirmness. It is light-demanding; cultural methods are similar to sessile oak.

Quercus ilex

HOLM OAK

Native of southern Europe and North Africa, introduced to England probably before 1500. This is an evergreen (*ilex* is Latin for 'holly'). The leaf is simple, very variable in shape and not lobed like other oaks in Britain, although sometimes toothed. Leaves are woolly when young, later green above and yellow-grey below.

The tree is not grown in Britain for its timber (although it is of good quality) but is useful as an ornamental and a windbreak on light, sandy soils. Avoid lime and clays. It tolerates salt winds. Although a southern species, there are some large specimens as far north as the Moray Firth on light soils near the sea. It is at its best in south-west England and the Channel Islands.

RED OAK

Quercus rubra

Eastern North America; introduced to Britain in about 1739.

This is a fast-growing tree with a straight stem and spreading crown. It is easily recognised by its large, spike-lobed leaves which turn gleaming red in early autumn. The leaf-lobes (usually seven to 11) are drawn out to fine points. Twigs are shining red.

The acorns are fat (more than 1cm across) and set on almost flat cups. They take two years to ripen.

Silviculturally the tree is not exacting; it dislikes lime but grows strongly on acid sands. It is frost-hardy and relatively shade-tolerant. It coppices well and is a useful understorey in pine stands.

The timber has not been much used here, although of fair quality. The tree is grown for its colour, for shelter (windfirm and fast growth), as an understorey and as a broadleaved tree to mix with conifers on light soils and slag-heap reclamation.

Fagus sylvatica

BEECH

Native to the south of England but not to Ireland or Scotland. One of the most important broadleaves of Europe, in closed forest reaching over 40m, with long, branch-free boles. The bark is smooth and silver-grey. There are ten species of *Fagus* in the north temperate zone and several varieties, notably the copper beech.

Leaves are alternate: when young, covered with silky-soft hairs; later smooth, mid-green and shiny. The spreading crown casts heavy shade. The zig-zag twigs carry sharp-pointed, shiny-brown buds, 1cm long.

The fruit is a prickly, four-lobed cup with two three-sided nuts. Abundant seed crops ('mast' years) generally follow warm summers; good seed even in northern Scotland. The germinating seedling has two kidney-shaped cotyledons held horizontally.

Beech prefers lime-rich soils and a mild, moist climate. Unsuited to waterlogged clay soils and dry sands. Suffers from early or late frosts; strongly shade-tolerant. On fertile limestone soils beech attains 25–30m in height even at 300m elevation in the north of England. It coppices moderately but may be pollarded.

Although usually windfirm, beech has very shallow feeding roots, immediately below the leaf-litter and easily damaged by trampling, vehicles or paving. Old beech are sensitive to drought, especially if their roots have been maltreated.

Beech is a valuable furniture and carpentry timber widely used for parquet flooring, veneering and food-contact items. An outstanding park tree and hedging plant.

HORNBEAM

Carpinus betulus

Native of south-east England and all Continental Europe except Spain, Portugal and Scandinavia. A natural associate of pedunculate oak.

A moderate-sized tree which resembles the beech in winter except that the bole is fluted and ridged. The bark is smooth and pale grey.

The leaves are sharply and doubly saw-toothed, rather dull-green above and paler below. The male and female flowers form separate catkins on the same tree. The twigs are zig-zag in shape; the buds (alternate) are slender and pointed, closely pressed to the twigs.

The fruit is a ribbed nut fixed to a large three-pronged wing; ten to 20 or more of the fruits hang in a tight cluster.

The timber is white, tough, hard and heavy, sometimes used for tool handles, but not highly valued. Formerly the foliage was used as cattle food.

Hornbeam adapts to a wide range of sites, including heavy clays. It is frost-hardy and may be planted in frost-hollows where beech would be damaged. It is shade-tolerant, coppices very strongly and is an outstandingly successful hedging plant.

Betula pendula

SILVER BIRCH

Native of Great Britain, Ireland and all the rest of Europe except the Mediterranean coasts. The genus contains 40 species in the north temperate and arctic zones.

A graceful tall tree, up to 25m, with pendulous twigs. The bark of the stem is shining white; in age it changes sharply at the base to deeply fissured, black and iron-hard. Relatively short-lived, 60 to 90 years.

The leaves are typically diamond-shaped, with long tips and deeply serrated margins. The thin, flexible twigs are covered with pale grey resin warts; the young branches are shining smooth. The buds are tiny and sticky.

The flowers are in long catkins, male and female. The seed is a tiny winged nut, the wings each wider than the nut and rounded. The side lobes of the catkin scales are curved as in a fleur-de-lys.

This is a pioneer tree and a strong light-demander. It will grow on poor dry soils but is much less tolerant of soil wetness than downy birch. It is a regular associate of pine in Caledonian pine forest, especially in the eastern Highlands.

In Scandinavia it is an important timber tree, for plywood, furniture, parquet flooring and turnery, as well as for paper-pulp and firewood. It is a favourite tree for gardens and parks. In spring the sugary sap may be tapped and fermented to make birch wine.

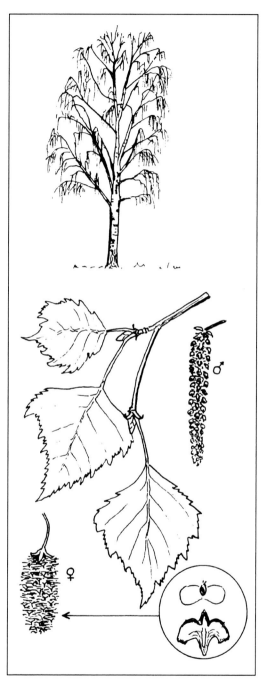

DOWNY BIRCH

Betula pubescens

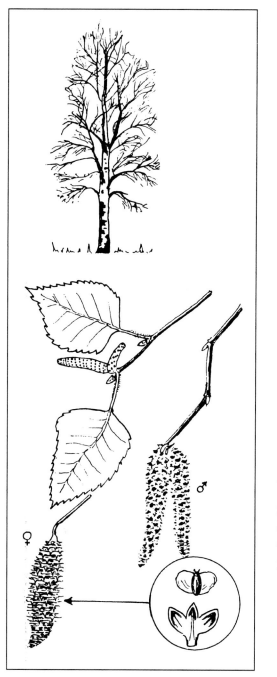

Native of Great Britain, Ireland, northern Europe and north Asia. It is variable in its features and has been divided into two subspecies, ssp *pubescens* in the lowlands and ssp *odorata*, a less erect form, in the uplands.

A small graceful tree with an upright, generally white bole. The branches, almost black, are spreading and usually ascending, rarely weeping.

The twigs are hairy, sometimes with small warts, but not rough. The leaves are usually wedge-shaped at the base, more bluntly pointed than silver birch and only slightly serrated.

The flowers are long catkins and the seed is a tiny winged nut. The wings of the seed are narrow and rectangular, and the catkin scale is *not* a 'fleur-de-lys'.

The tree is an important pioneer, tolerant of poor soils and harsh environments, and, especially ssp *odorata*, endures waterlogged and peaty conditions better than silver birch. It is slightly more shade-tolerant and grows as an understorey in open pine stands. Both birches may be used as nurses.

In Scandinavia birch timber is used for plywood, but in the UK its use is generally restricted to turnery, particle-board and firewood. It is, however, an important constituent of pine forest, useful for landscape design and important as a soil-improving pioneer of poor mineral soils.

Corylus avellana

HAZEL

Native of all Europe, including Great Britain and Ireland, and western Asia.

This is a large bush, occasionally a small tree, often forming an understorey in broadleaved woodland and woodland edges, but also extensive pure woodland on some exposed Atlantic coasts.

The leaves are short-stalked and double-toothed. The leaf-blades, stalks and twigs are densely hairy and the leaves are soft in texture, 5–10cm in length, almost as wide and coming to a sharp point.

The male flowers are catkins 5–10cm long; the female flowers are swollen buds with fine, bright red threads (the stigma) projecting from the points. The fruit is the well-known hazelnut which is contained in a ragged leafy cup, often in pairs.

Hazel grows on a wide range of soils but is best on deep, nitrogen-rich loam in a sheltered, sunny position.

The bush has been widely cultivated for its valuable nuts. It is shade-tolerant and coppices well. It was formerly used for making hurdles. It is the earliest source of pollen for bees.

Coppice management appears to suit this tree particularly well; without periodic cutting, the plants tend to become straggly and weak, and then die out.

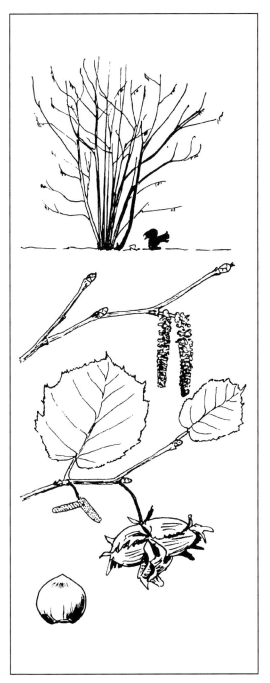

WILD APPLE

Malus sylvestris

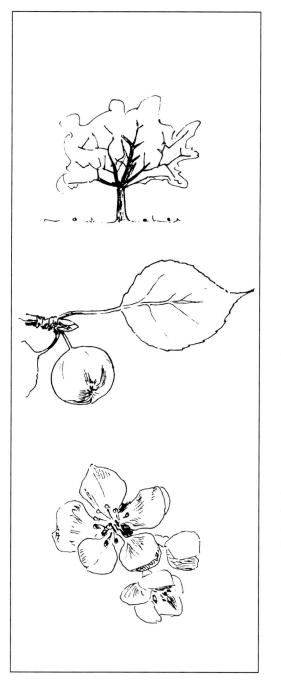

Native of the whole of Britain, all Europe and Asia Minor. This is a small tree or large bush, up to about 7m in height.

Twigs are red-brown, with many short shoots. Buds are alternate. Leaves are 3–4cm long, and oval in shape. The blossom is white tinged with pink, and showy. The fruit is a small apple, 2–3cm across, yellow-green in colour, often with a red flush.

This tree belongs to open broadleaved woods and scrub rather than forest. It is best on moist loam soils and in mild climates; it is much less common in the north of Scotland than in south-west England, Wales and Ireland.

Apple wood is tough, hard and heavy, formerly used for wooden screws, now for carving and mallets. It has a reddish-brown heartwood. The fruit is important food for wildlife.

A tree for amenity and wildlife conservation rather than commerce.

Various crab apples, including the Siberian crab, *Malus baccata*, and the wild pear, *Pyrus communis*, are not native to Britain. Some are recent introductions, some are garden hybrids and others may be descendants of cultivated trees brought to Britain by early Man. Pear wood is pink in colour and valuable, cutting cleanly and finishing well; it is used for turnery, musical instruments, etc.

Alnus glutinosa

COMMON ALDER

Native to all of Britain, the whole of Europe to the Caucasus and Asia Minor, especially on streamsides and on boggy plains. It may reach 33m in height, the bole extending to the tree top in the open crown. It coppices well and grows rapidly when young.

A micro-organism, *Frankia*, forms bright yellow nodules on the fine feeding roots, fixing nitrogen in the air, which it shares with the tree; thus the tree can thrive on nitrogen-poor soils and can improve their fertility, benefiting neighbouring trees.

The buds are stalked and blunt-ended with only two bud scales, the stalks triangular in section. The twigs, which are sticky when young, have resinous warts; buds and twigs have a waxy, bluish bloom. The leaves are circular, 7cm across, with a wedge-shaped base. They are shining dark green above, lighter below, turning black in autumn.

Male and female flowers appear on the same trees, the male 5–10cm long and rich purple in colour. The green female 'cones', 1–1.5cm long, grow in clusters. They persist, woody and black.

It grows on a wide range of soils but not on acid peat or where water lacks oxygen. It is an excellent nurse for spruce and oak, it improves fertility and its stiff twigs do not 'thresh' neighbours (cf. birch).

The wood is even-textured, pale when cut, rapidly turns orange on exposure and then fades to pale brown. It is good for cabinet-making, turnery and carving, known formerly as 'Scotch mahogany'. It is durable in water and makes excellent charcoal.

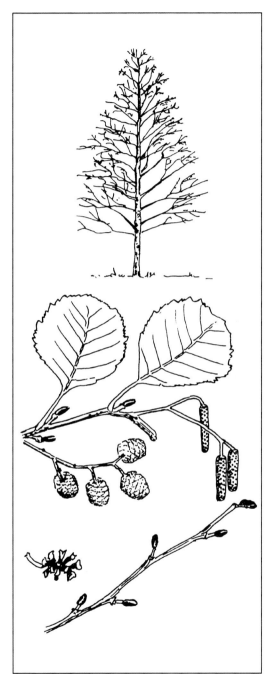

GREY ALDER   *Alnus incana*

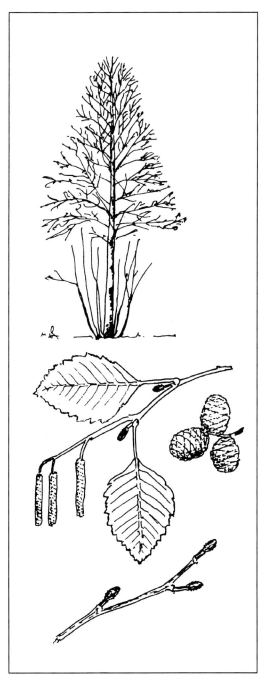

Native to Scandinavia, Central and Eastern Europe and the Caucasus, it was introduced to Britain some time before 1800 and is naturalised in Scotland. It grows to about 20m in height, the stem sometimes misshapen. It coppices well and suckers freely.

It may be distinguished from the common alder by the leaves which are oval with a sharp point. The upper surface is dark green with fine hairs, the lower side pale bluish-green with soft hairs. The mid-rib and lateral veins (about ten pairs) are prominent on the lower side of the leaf; each lateral vein runs right to the margin at a toothed lobe. The buds are stalked, hairy but not sticky, and blunt at the tip.

The 'cones' are a little smaller than those of the common alder.

It can be distinguished from red alder by the blunt-tipped buds, the smaller male catkins ('tassels') and the smaller cones.

Grey alder is lime-loving, fully frost-hardy and tolerant in respect of soil. It is a pioneer with a strong root system. Used on industrial wasteland. This is a much underrated tree.

The timber is similar to common alder but somewhat inferior.

Alnus rubra

RED ALDER

A native of western North America, from Alaska to northern California, introduced to Britain only about 1900. Also known as Oregon alder. In its home it grows on wet soils in association with Sitka spruce, and trials in Britain were to provide an associate for Sitka on peat sites. These were unsuccessful but showed it grows here on heavy clays and, thanks to its nitrogen-fixing micro-organism, on infertile industrial wastelands; these are its main uses in Great Britain. It grows very fast in youth (in excess of 1m a year on suitable sites). Although sprouting well if cut over up to about four years, it coppices poorly in later ages.

Red alder is similar in leaf shape to grey alder but in spring the flushing buds and young leaves are bright pink – hence the name. As in all the alders, the buds are stalked but these are sharp-pointed. The male catkins are much longer than those of grey alder and common alder, 10–15cm long, and the 'cones' are larger too, 1.5–3cm.

It is promising for farm woodland and land reclamation but it is important to get seed of a suitable origin: Alaskan is too slow-growing and Oregon and Californian are liable to frost damage. Seed from the Vancouver area is suitable, as is that from good Scottish stands in Lennox Forest at Glasgow. Production of small roundwood for particle-board on a rotation of about 15 years is a possibility being investigated.

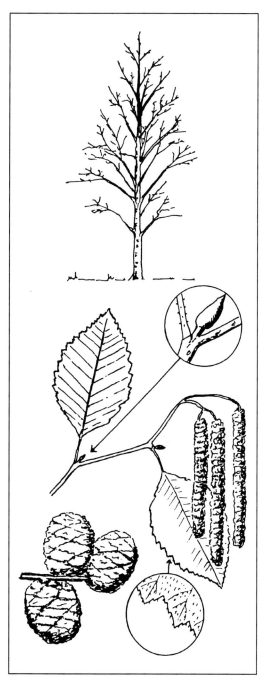

CORDATE ALDER

Alnus cordata

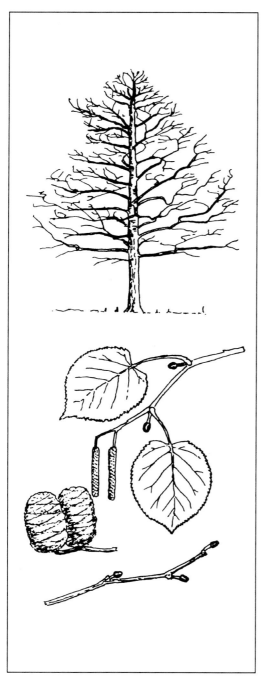

Native of Corsica and southern Italy, where it may reach 25m in height. It is also known as Italian alder.

The leaves are heart-shaped, bright shining green and large, up to 10cm long and 8cm broad. The lower surface is light green. The buds are on long stalks.

The 'cones' are larger than those of both common alder and grey alder, 2–3cm long. The bark of the bole is generally a greenish-grey.

The tree is somewhat susceptible to spring frosts but is a useful pioneer for planting on very base-rich soils, especially in the south and west, although it grows quite well as far north as Inverness. It tolerates slight shade and coppices well. A very vigorous and generally hardy tree, worthy of wider use. Like the other alders, this tree has an associated micro-organism fixing nitrogen.

Acer pseudoplatanus

SYCAMORE

Central European mountains; Asia Minor; introduced in Scotland before 1500. Fully naturalised in the UK. There are about 150 species of Acer, many ornamental.

A large tree with a dense spreading crown, reaching 35m. The five-lobed leaf is normally dark green above and either pale blue-green or, in one variety, pale purple below. The red leaf-stalk does not contain white juice. Black spots on the leaves, at first yellow, are a good identification, caused by a fungus, *Rhytisma acerinum*, which does no harm. The cleaner the air, the more common the spots. In heavy pollution, the leaves are clear.

The buds are green and opposite.

The fruit is the two-winged 'aeroplane propeller' with the blades set roughly at a right angle.

Sycamore likes moist air and deep mineral soil. It is frost-hardy and fully wind-resistant, and able to grow a full crown in severe exposure. It is very important as a shelterbelt tree, even in coastal areas. It coppices very strongly and grows fast on good soils.

The timber is highly valued for cabinet work and turnery, fine-grained and finishes well. Ripple-grain wood is especially valuable.

Although previously denigrated because of its invasive seeding in some ancient native woodlands, sycamore is now regarded more favourably by conservationists. It is of great importance for commercial timber-growing and as a component of farm woods. Liable to severe damage by grey squirrels. Forms a fine parkland tree to advanced ages.

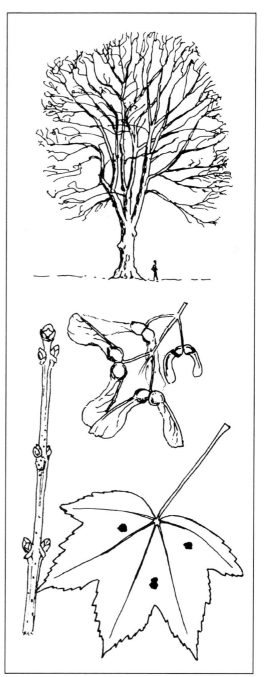

NORWAY MAPLE

Acer platanoides

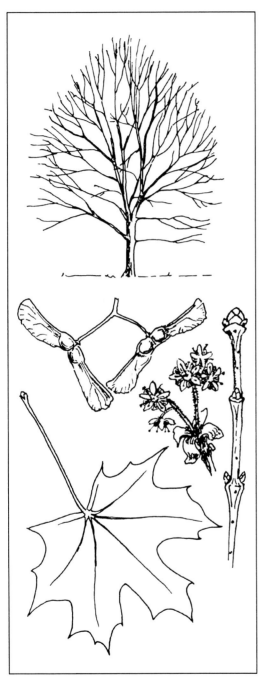

Native of Central Europe, including southern Scandinavia, Asia Minor and the Caucasus; first introduced in Britain, probably to Scotland, around 1650.

The middle-sized tree grows quickly in youth; more amenity than for timber.

As all other Acers, it has opposite leaves, buds and branching. Easily distinguished from sycamore by the leaves, which have sharp-pointed lobes and rounded indentations, and by the leaf-stalks exuding white juice when broken. No black spots on the leaves (cf. sycamore). In autumn the leaves turn brilliant yellow and gold, a major asset in community and amenity woodland.

The blades of the 'propeller' fruit are set at a wide angle, half-way between a right angle and a straight line.

The winter buds are red.

Norway maple is semi-tolerant of shade, more adaptable than sycamore but thriving on similar sites. It is windfirm and useful for shelter planting on the farm. Cultivation is similar to sycamore but it is less invasive. An excellent park tree.

The timber is similar to sycamore's and finishes well for cabinet-making.

The flowers appear before the leaves and are a valuable source of honey for bees; in June bees also take leaf-nectar.

Acer campestre

FIELD MAPLE

Native in England and (perhaps) southern Scotland, as well as most of Europe and Asia Minor. Not native in Ireland.

A small tree of hedgerows and open woods. It is essentially a lowland tree on fertile, lime-rich soils.

It can be distinguished from sycamore and Norway maple by the smaller size of all its parts. The twigs often grow corky ridges or wings along their length. Buds opposite.

Leaves are opposite, dull green above, paler below, about 7cm across. Buds and leaf-stalks exude milky juice when cut.

The fruit is a two-bladed 'propeller' with the blades forming a straight line (occasionally even slightly bent backwards).

The timber is of no commercial importance but is tough and very fine-grained. It is valued for turnery and was formerly used for making mazers, ancient wooden drinking cups. The foliage is eagerly taken by stock and deer. The flowers attract many bees.

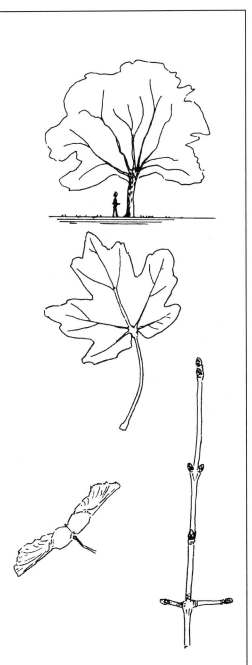

LONDON PLANE

Platanus acerifolia

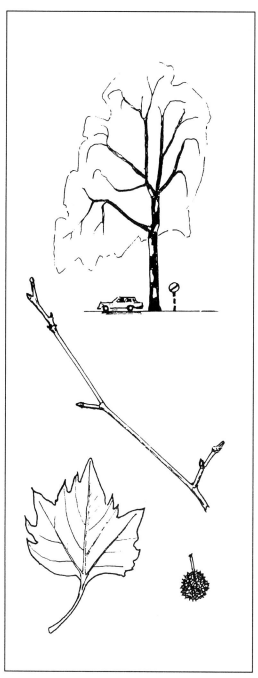

This is a natural hybrid between *Platanus occidentalis* and *P. orientalis* which appears to have originated in Oxford in about 1670. It is essentially a town tree, now commonly planted in cities worldwide. It often grows to over 30m in height. *P. orientalis* is a native of south-east Europe and Asia Minor, introduced to England before 1600. *P. occidentalis* is from eastern North America, brought to England in 1637. Neither grows as well here, or in cities elsewhere, as the London plane.

Although the leaves are shaped like the maples', it is easily distinguished from them by the alternate buds and leaves, the stalked buds and the spherical fruit. The fruit is a spiky ball, about 4cm in diameter, breaking up in spring and leaving the fruiting stalk persistent on the tree.

The spiky, leathery leaves are glossy, harsh green above, paler below, and fall in early November without colouring. The hollow bases of the leaf-stalks completely enclose the winter buds.

The bark is a marked feature; it is dull grey, flaking off in patches each summer to show the yellow under-bark.

Although slightly frost-tender, the tree tolerates poor soil, mutilation and heavy air pollution. It prefers soils which are not lime-rich.

The wood is attractive, with a yellow-pink colour, rather like beech. It is marketed as 'lacewood' and used for furniture and ornamental purposes.

Aesculus hippocastanum

HORSE CHESTNUT

Native to small areas in the Balkans and Caucasus and introduced to Britain about 1670. A large, handsome tree, up to 30m in height.

The well-known compound palmate leaves usually have five to seven leaflets. The buds are large, fat, very sticky and opposite. The leaf-scars are like horseshoes, even to the nail-holes.

Flowers are showy white 'candles', usually very prolific. The fruits are more or less prickly, fleshy capsules, each with three sections but often containing only one large shiny seed – the well-known 'conker', not good for human consumption but eaten by animals.

The wood is soft and white, not very valuable. It is easily split. The tree is valued for amenity, thriving on very base-rich soils (including chalk and limestones), and will grow on heavy clays, where it is a useful associate of oak. It casts heavy shade and provides a heavy litter. It tolerates some shade. Plants are easily grown from seed. Prone to stem rot after poor pruning.

The red-flowered *Aesculus x carnea* is a cross between *A. hippocastanum* and *A. pavia*, the red buckeye of the southern USA. It is generally smaller than the white horse chestnut, with deep rose-coloured flowers. It is much more frost-tender than the white, largely confined to the Midlands and south of England, where it produces fertile seed. Although seedlings grow, good red varieties are best produced by bud-grafting. Not a long-lived tree.

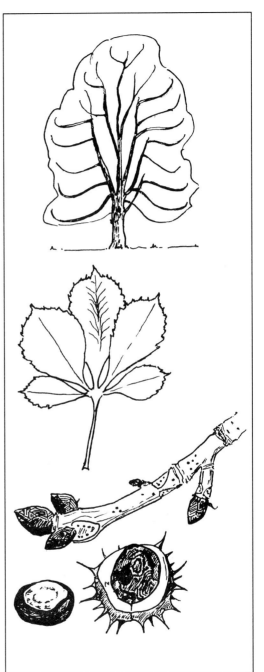

SWEET CHESTNUT

Castanea sativa

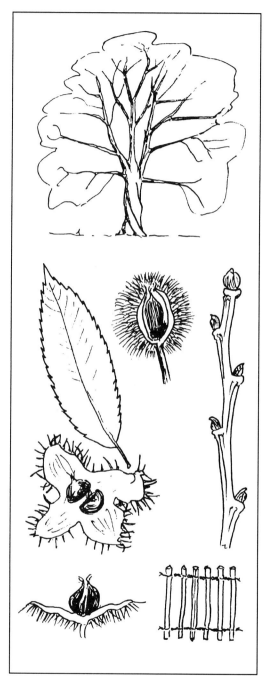

Native of southern Europe and Asia Minor, introduced to England before the Norman Conquest. It is a large tree with a massive bole which quite often shows spiral growth. The buds are alternate, yellowish-green and downy. The leaves are simple, spear-shaped and large (10–25cm long); they are prominently toothed and deep green in colour, changing in autumn to rich dull brown.

The very spiny capsule contains, usually, two nuts, each a complete fruit containing one seed (cf. the horse chestnut's spiny fruits containing conkers which are truly seeds). Although the tree can grow to a large size on suitable sites even in the north of Scotland, seeds are fertile only in the south of England. Most edible nuts are imported from Italy.

The tree produces a high-quality timber, very similar to oak, which is strong, durable and finishes well. It cleaves very well and smaller sizes are used to make cleft-chestnut fencing.

The tree is frost-sensitive and grows best in mild, moist climates. It is moderately shade-tolerant but does best in full light, on moist, well-drained, fertile soils. It grows badly on lime-rich soils, which should be avoided. It coppices superbly and is by far the most common species in actively managed coppice woodland in Britain.

Juglans regia

WALNUT

Native of south-east Europe, Asia Minor and the Caucasus to the Himalaya and China. It has been cultivated in Britain since very early times, perhaps pre-Roman. The name, from Old English *wealh*, means 'foreign nut' (cf. hazel, the native nut).

A large, handsome tree with alternate buds and compound, pinnate leaves, normally with seven leaflets. Buds, leaves and flowers, when crushed, are highly aromatic. All kinds of walnut can be recognised by the chambered pith of the twigs.

The male flowers are catkins, purple when ripe. The small, erect female flowers are in clusters of three, developing into a fleshy and spotted green fruit containing the well-known walnut. Rooks and squirrels greatly enjoy the nuts.

The timber is highly valued for all cabinet-making, turnery and carving. The burrs are especially valuable for veneers. The tree is particularly vulnerable to stem-rot with advancing age; for that reason special care should be taken in pruning to produce a clean finish and any dying side branches pruned before rot spreads downwards.

The tree should be grown only on highly fertile, deep, moist soils in the warm lowlands. It is light-demanding and frost-tender, and should generally be planted in a nursing mixture, often with ash. Easily grown from seed but plants develop a strong tap-root which makes transplanting difficult. They coppice well. Country lore says: 'The dog, the wife and the walnut tree, The more you beat 'em the better they be!'

The related *Juglans nigra*, the black walnut from the eastern USA, may be planted as an ornamental; it has more leaflets, usually 12 or more. The timber is equally valuable and the nuts similar. Culture is similar to the European walnut.

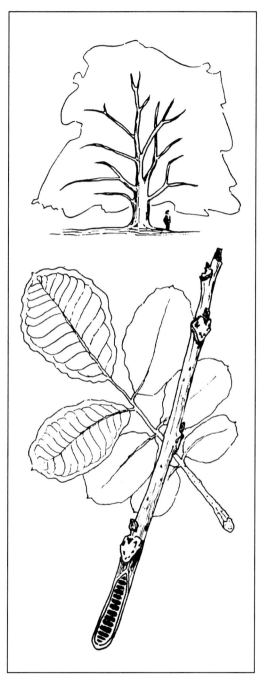

ASH

Fraxinus excelsior

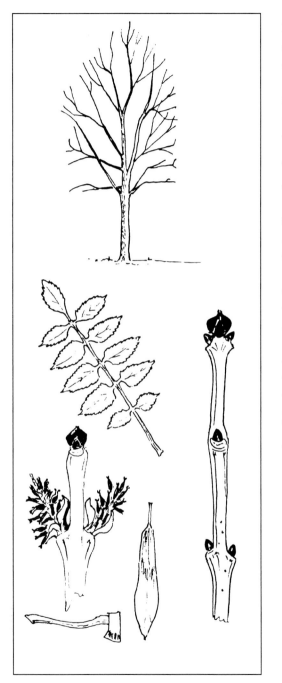

Native to all Britain and Continental Europe except southern Spain and northern Scandinavia; also to Asia Minor and Russia.

Ash may grow to 35m in height in favourable places, with a long, clean bole and an open, arching crown. The bark is grey, sometimes with a golden tinge, forming rhomboidal shapes between deep fissures. Twigs are pale olive-grey. The leaves are opposite, compound, with a mid-rib 15–25cm long, three to six pairs of leaflets and a terminal one, each 5–10cm long and irregularly toothed. The first leaves fall in autumn, never colouring.

The buds are velvet black, horny and opposite. The flowers are purple and the fruits or 'keys' are 'single-bladed propellers' hanging in tight bunches on the tree right into winter.

Ash may appear in places where it will not succeed to maturity. It is light-demanding except in youth, favouring deep rich mineral soils. It grows on moist soils in the lowlands and on moist scree slopes on very fertile rocks. The water must not be stagnant. An exacting species which grows fast and coppices well.

Well-grown timber (wide, even growth rings with a white or pinky colour) is resilient and valuable for sports goods, tool handles and cabinet-making. Slow-grown wood ('brown ash') lacks resilience.

The ash plays a vital role in Norse mythology: it is Yggdrasill which, with its roots and branches, binds together hell, earth and heaven.

Tilia cordata

SMALL-LEAVED LIME

Native of England and Wales, northern and Central Europe and Russia, in mixed, broadleaved woods. A tall and handsome tree (up to 25m), it is thickly leaved and casts a heavy shade. A soil-improver.

The zig-zag twigs are characteristic due to the terminal bud continually aborting and a lateral one developing. Twigs are shining red on top, olive green below. The buds are also red on the side receiving most light, stout and rounded, each with only two or three bud scales. The leaves are heart-shaped with toothed edges, rarely as much as 7cm long on 3cm stalks. They are fine-textured, deep green and shiny above, very pale green below.

The flowers open in late July, pale ivory or pale green in colour, very sweet-smelling and in clusters of four to eight. Each flower grows on a long stalk arising from a wing-like leaf which helps dispersal by wind. The fruit is a small, fragile sphere attached to the wing, but it sets good seed regularly only in the warmer parts of England (roughly south of York).

Planting of lime in parks and woods is usually the common lime, and the small-leaved lime is rather neglected. It is most common in the west and south-west of England, especially in limestone areas. It coppices well and grows as an understorey in some oak woods, tolerating some shade. Plants should be raised from seed, not taken as cuttings or layers.

The wood is white, useful for carving, turnery, drawing-boards, chests, musical instruments, etc.

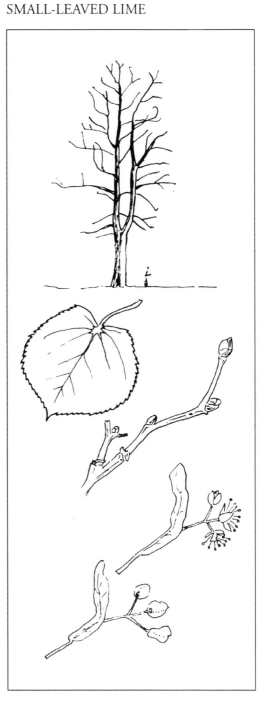

COMMON LIME

Tilia x vulgaris

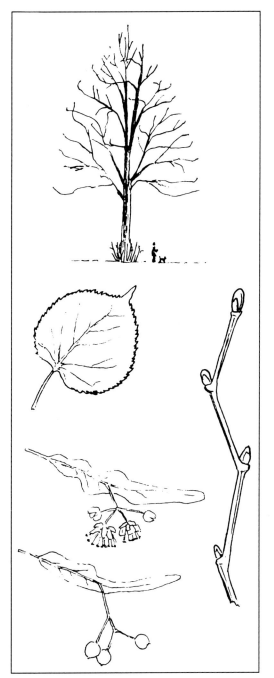

During a warm climatic period some 7000 years ago, the large-leaved lime, *Tilia platyphyllos*, grew over much of northern Europe, including Britain, but it is now restricted to southern Europe and, even planted, is rare in Britain. Its leaves have blades 5–12cm long and, in contrast to *T. cordata*, the buds, young shoots and leaves are downy.

The small-leaved and large-leaved lime interbred spontaneously to produce the hybrid *Tilia x vulgaris,* now the lime most commonly planted in Britain. It resembles the small-leaved lime except that the undersides of the leaves are pure green and 6–10cm long on normal shoots (maybe longer on suckers). The leaves and shoots are hairless. In favourable situations it grows up to 40m.

The flowers are strongly scented with much nectar, and are important for bee-keeping. Seed is produced in most years in the warmer parts of the country, seldom in Scotland although the tree grows well there.

The tree is hardy, windfirm and important for shelter and ornamental planting. There are many handsome avenues of this tree. It coppices well and is easily reproduced vegetatively, by layering (unfortunately the trees which are easiest to layer are those with heavy sucker growth at the base, and this unsightly feature may be perpetuated by this means). Plants raised from seed are greatly to be preferred.

On fertile brown-earth soils and in shelter, this tree grows quickly. A good street tree (though car owners dislike the dripping nectar) and ornamental. An excellent soil-improver. The wood is white, turning pale brown, lightweight and even-textured; it cuts crisply for carving and is used for turnery.

Prunus avium

GEAN

Native to Great Britain and Ireland, all Continental Europe, Asia Minor and western Asia. The natural range was probably greatly extended by Man, even in pre-history, so it may not be truly native to much of Britain.

A middle-sized tree, occasionally reaching 25–30m. In young trees the form is very regular with whorled branches, but it becomes ragged and rough with age. The branches are generally ascending. A notable feature is the horizontal stripes on the bark, which are corky lenticels.

The leaves are elliptical, doubly serrate with a sharp tip. At the junction of the leaf-stalk and the blade are two glands, red or sometimes yellow. Foliage usually turns scarlet in autumn.

The flowers are showy white blossoms with five petals, opening usually in May as the leaves appear. The fruit is a blackish-red cherry with a smooth, spherical stone, greedily eaten by birds.

The young plants are fairly shade-tolerant but become light-demanding with age. The tree grows well on moist fertile soils, especially with lime. It is a natural associate of oak on heavy soils. It grows quickly and can produce an early return on good sites. The timber is valuable for furniture, veneers and turnery. It is a promising tree for planting on fertile farmland and in mixed woods, good for its blossom, its autumn colour, for wildlife and for its timber value.

The plants are used widely as rootstocks for grafting cultivated flowering cherries.

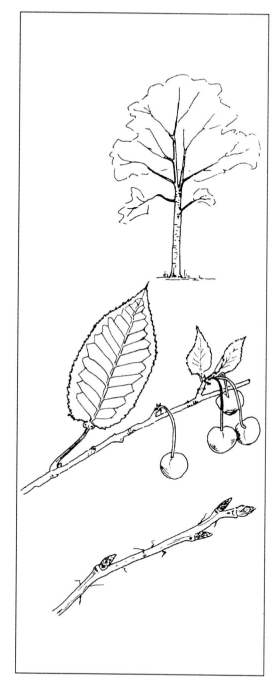

BIRD CHERRY

Prunus padus

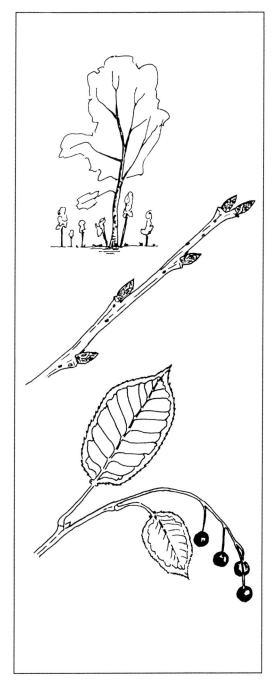

Native of Europe and Asia, including Great Britain and Ireland, typical of woodland edges and moist valleys. A modest tree up to 12m. It is more common in northern Scotland than in lowland England.

The stem is often much divided, with numerous root suckers. The bark is thin, with shallow furrows; it is typically dark grey but seems very attractive to lichens and algae so is often a patchwork of dark purply-grey and light greeny-grey.

Buds are slender and pointed. The leaves are dull green, wrinkled and finely toothed. Each leaf has two glands at the top of the leaf-stalk. Each lateral vein on the leaf-blade runs forward to join the next vein up the leaf to form a rather obvious marginal vein.

The blossom, generally in late May, is white, strongly scented and attractive to bees. The fruits are small cherries, bitter-tasting but very attractive to birds; the small stones are oval, pointed and with a crest.

The tree is quite demanding, thriving on moist loams; it coppices strongly and grows rapidly when young.

The bark contains amygdalin (prussic acid) and is mildly poisonous. Since the bole is small, the timber is not greatly valued but the heartwood is an attractive golden-brown, used for carving and turnery. The tree stands exposure well and is useful for shelterbelts; it is attractive for landscape planting on woodland edges and for the support of many birds.

Sorbus aucuparia

ROWAN

Native to the whole of Britain, all Europe, Asia Minor and Siberia. A middle-sized tree, seldom attaining 15m, with a slender stem and a very open crown. The bark is smooth, golden-grey and glistening, with rings of lenticels.

There are long and short shoots and the buds are large, alternate, spirally arranged and downy. The leaves are compound and pinnate with usually five to seven pairs of leaflets plus a terminal one of the same size. All have sharply toothed edges. In autumn the leaves often turn bright orange and red. (The name mountain ash, based on the superficial similarity of the leaves, is confusing since the two trees are quite unrelated. Rowan is the Old English name.)

The flowers are creamy white, in flat-topped bunches, opening in May or June, and have a heavy smell. The fleshy fruit is a brilliant scarlet 'berry', round and pea-sized, ripe in September. They are eaten by birds, which spread the seed, and they make excellent jelly to eat with meat. In most years there is a heavy crop.

Rowan is very undemanding, growing on thin dry soils, heathlands and peat bogs and the limit of woodland. It is frost-hardy, light-demanding and excellent for woodland edges and the margin of shelterbelts. Not very long-lived; with age it tends to fall apart. The wood is dense, pinky-brown in colour and used for turnery, marquetry, carving and tool handles.

A park tree and a useful street tree. Important in supporting berry-eating birds. There are about 90 species of Sorbus in the northern hemisphere, including many ornamentals from China, which are closely related to rowan; many garden cultivars have been developed with variously coloured flowers and berries.

Throughout Gaeldom the rowan is held to have magical powers, especially to counter witchcraft.

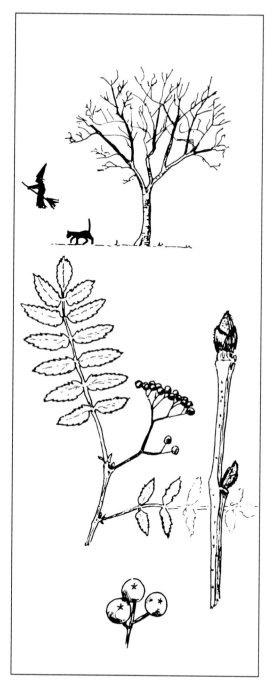

WHITEBEAM

Sorbus aria

Native to England, Wales and much of Europe and North Africa. Rare in Ireland. It favours limestone soils and grows to a middle-sized tree, up to 20m, in open woodland and on forest edges. It is planted and grows well in Scotland but may not be native.

The buds are alternate, spirally arranged, greenish-brown in colour and pointed. The leaves are simple and toothed (usually two sizes of teeth, alternately); the upper side is dull green and the lower side covered in pure white felt. The young shoots too are covered in thick white down.

The flower heads are large and showy, the flower stalks with white down and the flowers white, sweetly scented and attracting honey-bees.

The fruits are similar to rowan but some are slightly elongated and deep scarlet.

This is one of the few trees which succeeds on chalk, where it is a good associate for beech. It is an excellent tree for shelterbelt planting, even in very exposed places. Very hardy and, like rowan, an important tree for berry-eating birds. It is excellent for ornamental planting and is easily raised from seed.

The wood is used for turnery and tool handles; it is hard and finishes well.

Beam is Old English for 'tree', so whitebeam means 'white tree' – presumably referring to its foliage – meaning that 'whitebeam tree' is a tautology.

Sorbus intermedia

SWEDISH WHITEBEAM

Native of south Sweden, the Baltic states and islands, and north-east Germany; introduced to Britain in the seventeenth century.

The leaf shape is intermediate between the rowan and the whitebeam and it is one of several such intermediate forms and species, e.g. *Sorbus arranensis*, the Arran whitebeam (confined to that Scottish island), and *S. hybrida*, the Finnish whitebeam, both very similar to the Swedish.

The Swedish whitebeam is more compact than rowan, with a denser crown. The leaf has one to five pairs of lobes, with the rest of the leaf, towards the tip, entire and coarsely toothed. The underside is covered with yellowish-grey felt.

The flowers are similar in structure to the rowan's, although the clusters are heavier and the colour is creamy, instead of white. The fruits are barrel-shaped, longer than broad and coloured weak orange (rowan's are scarlet).

This is an important tree for the edges of shelterbelts and new woodland, for ornamental planting and for bird-protection areas. It is resistant to smoke pollution and very hardy, suitable for a wide range of soils.

SWEDISH

FINNISH

87

WILD SERVICE TREE

Sorbus torminalis

Native to middle England southwards and most of Europe to the Caucasus. Very local to rare distribution.

It forms a small tree up to about 20m in height, with an oval crown and ascending branches.

The leaves are bright green above, yellowish on the underside with down only when they are recently opened. They are symmetrically lobed, the lowest pair of lobes more deeply cut than the rest, and all the lobes are sharply pointed. In autumn the foliage turns bright red.

The flowers are white, similar to other Sorbus species. The fruits are small and barrel-shaped, similar to Swedish whitebeam but dull green when ripe. Much seed is infertile.

Buds are alternate, stout and green.

The timber is unimportant, although it is hard, durable and takes a fine polish. The tree is an associate of oak and ash on heavy clay soils in southern England and south-east Wales. The fruit was used medicinally in former times in the treatment of colic in man and in horses (*tormina* is Latin for 'colic').

Crataegus monogyna

HAWTHORN

Native to all Britain, Europe, Asia Minor and North Africa.

This is a large bush, sometimes a small tree, with a divided, often twisted stem and a thick, wide-spreading crown. It may attain 10m in height, occasionally more.

The buds are alternate and hairless. The twigs are also hairless and generally armed with many short spines; long spines tip the short shoots.

The irregularly shaped leaves are alternate, deeply lobed, dull green above and blue-green below. The lower lateral veins of the leaves may curve backwards.

The flowers are the well-known May blossom, pure white, very showy and profuse.

The fruit is the blood-red 'haw' berry and contains a single stone (*mono-gyna* means 'one-carpel').

The bush prefers lime-rich, humus-rich soils but it also grows on dry, sandy and stony land. It is an excellent hedge plant, producing coppice shoots. Extraordinarily hardy and wind-tolerant.

Hawthorn contains many volatile oils, tannic acid, glycoside, triterpene and other acids which are used in pharmacy and homoeopathy. Much superstition is attached to the tree. It is used principally as a hedge plant.

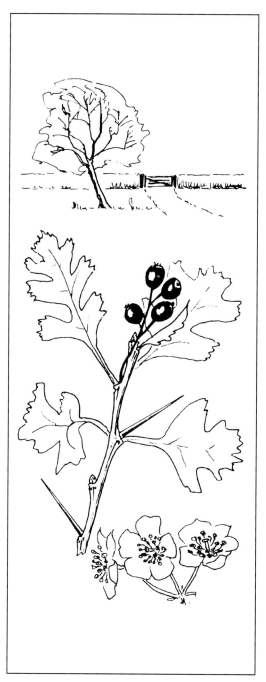

MIDLAND HAWTHORN

Crataegus laevigata

Native of the south of England and the rest of Europe. It grows in hedges, in scrub woodland and as an understorey tree; a slow-growing barrier shrub, occasionally a tree up to 8m tall.

One-year-old twigs are light olive-brown or reddish-brown. The buds are alternate.

Some leaves are shining dark green and may be lobed (less so than the common hawthorn) and some are not divided. Typically the lowest lateral veins point forwards towards the leaf-tip.

The white flowers, which normally appear about two weeks earlier than hawthorn's, have an unpleasant smell. They have two stigmas and two seeds per fruit (occasionally three).

The thorn coppices well and makes an excellent hedge.

The midland thorn will grow on a wide range of soils, including heavy clays. It is moderately shade-tolerant.

The thorn is planted as a hedge but less commonly than hawthorn. Also in bird-protection areas for its berries. The wood is hard and very heavy, used for turnery.

Most of the garden cultivars of thorns, especially those with red flowers, have been developed from the midland thorn. There are about 100 species in the genus *Crataegus*, mostly natives of North America.

Ilex aquifolium

HOLLY

Native to Great Britain (but not Ireland), western and southern Europe, North Africa and Asia Minor. Absent from most of Scandinavia, the Baltic and Russia.

This is an evergreen, an understorey bush or small tree in shady broadleaved and mixed woods. It has a bushy habit, responding well to coppicing and clipping. It is an excellent hedge plant. It can grow to 15m in height.

For the 2m or so close to ground level the leaves are armed with sharp spines, but above that the leathery deep-green leaves are simple and unarmed.

The male and female flowers grow on separate trees, attracting many bees. The small white flowers have a sweet smell; the fruits are the well-known gleaming red berries, eaten greedily by birds which disperse the seed.

The genus *Ilex* contains about 300 species, temperate and tropical.

Holly grows well on a wide range of soils, from acid to limy. It tolerates shade and is smoke-hardy. It is principally used as an ornamental but is an important component in understoreys for beech on limestone areas and oak in the west.

The wood is white and fine-textured, cuts cleanly and is used for carving, inlay work and turnery.

There are very many garden varieties.

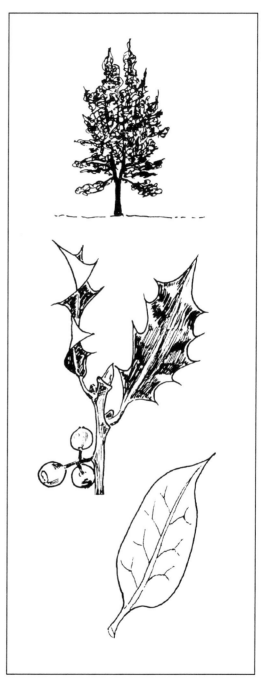

WYCH ELM

Ulmus glabra

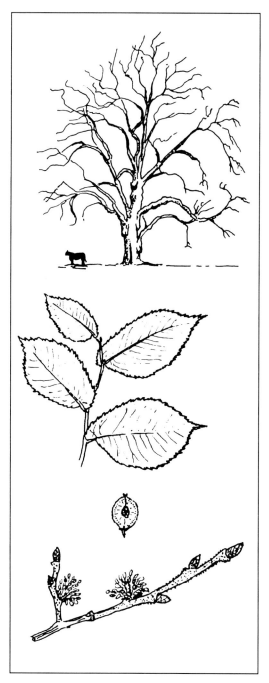

Native of all Britain, Europe and the mountains of Asia.

The elms can be recognised by their simple alternate leaves with unbalanced leaf-bases; the fruits are roughly circular wings with a single seed, a tiny nut, more or less in the centre of the wing.

The genus is susceptible to the Dutch elm disease, a fungus – spread by a common bark beetle – which grows into and blocks the vessels in the wood. This causes branches and eventually the whole tree to die. The common English elm showed virtually no resistance and has been almost wiped out except for young suckers in hedges. The wych elm has more genetic variation than the English and many remain alive, although trees are still becoming affected.

It is a large tree with a dense crown. The leaves, usually 8–10cm, have unequal sides at the base and the top surfaces are as rough as medium sandpaper.

The buds are alternate, sharp-pointed and hairy. The flower buds swell early in the year and blossom early (usually March) as pink globes; within three weeks the abundant green fruit wings hang in clusters before the leaf-buds break. The fruits are fully ripe and pale brown in late May or June and are released in great numbers. Seed sown green germinates immediately; sown brown it will germinate the following spring.

Wych elm thrives best on moist, base-rich light loams. It is windfirm and frost-hardy. The severity of the Dutch elm disease is such that it cannot be recommended for planting at present; control of the bark beetles which spread the fungus may allow future planting.

Elm timber is strong, has a beautiful figure and finishes well. It was used for chair-making, wheels, chests, coffins and heavy carpentry.

Salix species

WILLOWS

There are about 300 named species of the willow family, most from the temperate and arctic areas, but family members hybridise easily and the number is open to debate.

Most willows are deciduous; some are large trees, some bushes, some tiny, creeping plants. Most belong to open and damp situations; they reproduce easily from cuttings and sucker and coppice readily.

The flowers are catkins, male and female on separate trees. The male catkins of some are attractive 'pussy willows', harvested for florists. After fertilisation, the female catkin becomes swollen and covered in a mass of silky hairs which are attached to the minute seeds and serve as their parachutes for dispersal.

All willow seeds have a very short life; to survive they must land on moist ground and germinate immediately.

Willow buds are arranged alternately (really spirally) on the shoot and are very unusual in that they are protected each by a single bud scale.

Willow timber is lightweight and not durable, but most species are tough. One is cultivated especially for making cricket bats. Willows are grown widely for basket-making, notably *Salix viminalis*, the osier. Most willows grow rapidly, at least when young; some high-yield varieties are now bred as energy crops (see chapter 14).

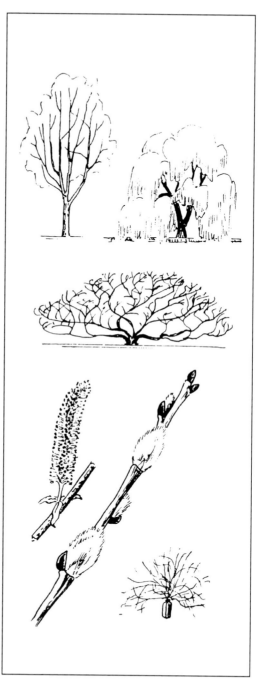

Top left: cricket-bat willow
Top right: weeping willow
Centre: grey willow
Below: catkins
Bottom right: willow seed

GOAT WILLOW

Salix caprea

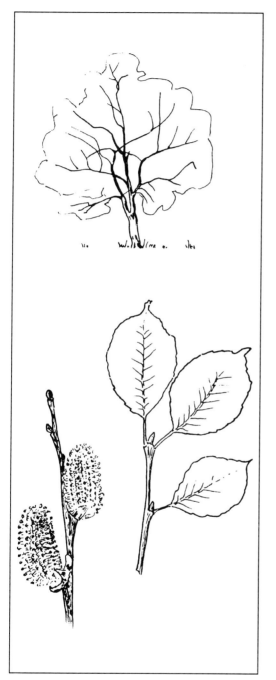

Native throughout Britain and also to all Continental Europe and Asia, as far as Korea. A small tree, up to 10m, growing in woodland and on open ground. Also known as sallow.

The leaves are oval to round, about 5cm, with the top side smooth and bright green, the lower side felted pale bluish-green. Twigs are downy in their first year, becoming smooth grey-brown.

The catkins come out early, before the leaves, usually in March. The male catkins are 2–3cm long and important for bee-keeping as they provide early pollen.

The bark contains sufficient tannin to have been useful formerly for leather tanning, and it also contains salicylic acid, from which aspirin is made. The wood is not commercially valuable, although attractively pink-coloured.

The tree grows rapidly for a few years on a great range of sites and is an important pioneer for the establishment of new woodland of more valuable species. It is a favourite food for deer.

Salix alba

WHITE WILLOW

Native of most of Britain (except the extreme north of Scotland), all Europe, North Africa and most of northern and central Asia. A handsome tree up to 25m in height, although often pollarded (that is, all the branches cut off at the base of the crown to encourage regrowth; like coppicing, but 2–3m above ground). The twigs are not brittle; in some varieties they are pendulous, in others erect.

The leaves are lance-shaped and 5–7cm long. The lower surfaces are densely covered in white down (hence the name), the upper surfaces also downy but less dense. Buds are pressed against the twig. Male and female catkins are about 3cm long.

White willow thrives on moist, deep loam soil, tolerating wetness. It is grown as a park tree and some varieties are used for heavy baskets. The boles may be used for the same purposes as poplar timber.

The species has several distinct varieties which have different habits and forms of twigs, although some authorities regard these as separate species or as hybrids with *S. alba* as one parent.

Cultivar 'Chermesina', coral-bark willow, has coral-red twigs, striking for landscape planting. 'Coerulea' is the cricket-bat willow, an upright form with purple twigs, grown on short rotations in southern England for the manufacture of bats. 'Tristis' is the common weeping willow in Britain.

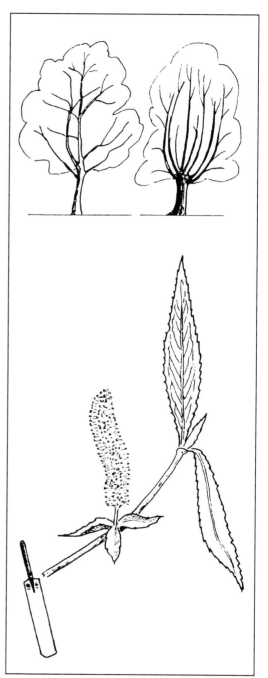

CRACK WILLOW

Salix fragilis

Native throughout Great Britain, over most of Continental Europe and east to Siberia and the Hindu Kush, but perhaps not in Ireland.

It may reach 18m in height, with a wide and open crown. Branches emerge at a wide angle to the stem, as much as 90°.

It is a waterside tree with brittle twigs. Leaves are lance-shaped and sharp-pointed with the margins coarsely toothed, each tooth pointed with a tiny red gland. The leaves are without hairs, mostly about 9cm long and 2cm wide. Male catkins are 3cm long and pointed.

The names '*crack*' and '*fragilis*' refer to the brittle wood. Twigs easily snap off where they join the branch and larger branches also fall, especially in spring.

Formerly the tree was often pollarded to provide poles for stakes, but the wood is not well regarded. Crack willow grows very readily from cuttings and it was used widely for drain-side planting to stabilise the banks.

Salix pentandra

BAY WILLOW

Native and common throughout Britain, extending over most of Europe and Siberia.

A small tree, 12m, with smooth shiny brown branches (as if varnished). The leaves are oval in shape, about 5–7cm long, finely toothed and similar in appearance to the bay leaves used in cooking (a similarity from which the common name is derived); they are shining, bright green and fragrant if crushed.

Male catkins are 4cm long, bright yellow and attract honey bees.

This is essentially a bush of the fens, on wet, fertile land.

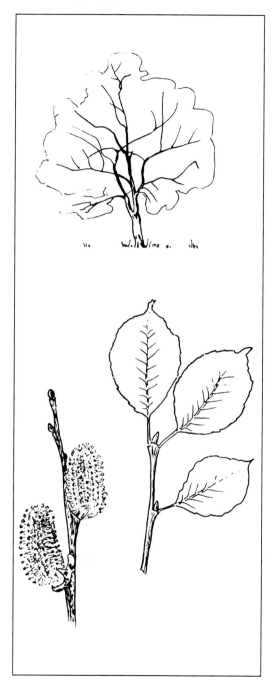

ASPEN

Populus tremula

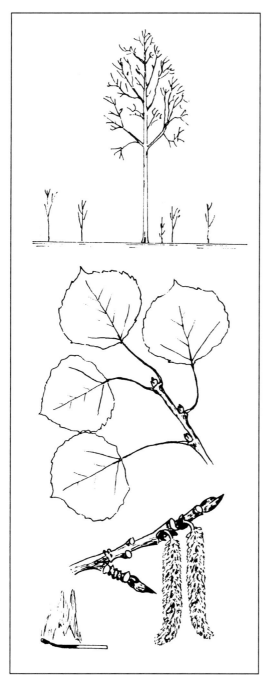

Native of all Britain, Europe, Siberia, Asia Minor and part of the North African coast, in the woodlands of the plains and of the mountains, often up to the top limit of shrubs. It may grow up to 35m in height. Aspen regenerates typically by root suckers. It has a rounded crown with the bole extending to the tree top. The bark is smooth golden-grey, cracked and speckled.

The leaves are almost circular, about 5cm across. They are held on long, flattened stalks which allow the leaves to flutter in the wind. In autumn they turn spectacular colours, from gold to deep purple.

Like all poplars, aspen has male and female flowers on separate trees. The male flowers are catkins about 6cm long, at first red then opening to grey-pink. The long female catkins develop into cottony tassels, each strand attached to a tiny seed. To survive, seeds must germinate immediately.

Aspen is winter-hardy and undemanding of climate. It requires some fertility and is best on moist loams. It is completely intolerant of shade and has a light, open crown. The rooting is shallow and it suckers freely. It grows rapidly in youth and makes an excellent nurse for frost-tender species.

The wood is white, soft and even-grained. It is used for match manufacture, artificial limbs, carving and veneering, including chip baskets.

The Gaelic name is *cran critheanach*, 'the shaking tree', and the Welsh *coed tafod merched*, 'the tree like a woman's tongue'.

Populus x canescens

GREY POPLAR

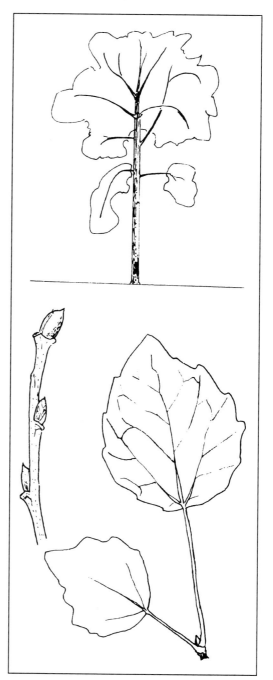

Native to Europe and Asia Minor, probably native to south-east England. It is a large, vigorous tree, up to 35m in height, easily recognised by the pale grey downy covering on the undersides of the leaves. The down is soon shed (unlike white poplar) when the undersides of the leaves are greenish-grey. The leaves on the long shoots and suckers are not palmately lobed (in contrast to the white poplar). The branches have smooth grey bark ringed with rows of black lenticels. The twigs are very downy in the first year, smooth later. Buds are golden and downy, not sticky.

The tree produces abundant suckers. It resists wind well and is excellent for shelterbelt planting. It is not sensitive to salt winds and tolerates smoke. It can grow well on calcareous soils.

Grey poplar is a handsome tree for ornamental parkland planting but is not suitable for small gardens. It is disease-free and propagates easily from cuttings.

This appears to be a hybrid between aspen and white poplar, i.e. *P. x canescens*. Importantly, it is canker-free.

WHITE POPLAR

Populus alba

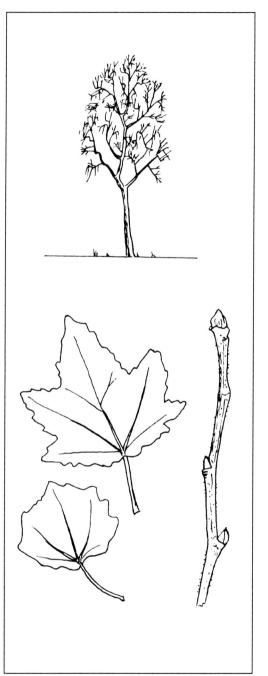

Native to central and southern Europe, Asia Minor, Russia and the Himalaya. It was introduced to England probably by 1550, although in early references it may have been confused with the grey poplar.

The buds and leaves have a white felty covering, later turning grey, the felt persisting until the leaves fall (this helps to distinguish this from grey poplar, most of whose leaves turn green). The leaf-stalks are round, not flattened.

The leaves on suckers and long shoots are palmately veined and palmately lobed, distinguishing it from the grey poplar.

The tree has no special value for its timber which is coarse and rough; its value is for shelter-planting and as an ornamental. A useful sea-side tree, it is best on moist, silty loams. A modest-sized tree, generally of poor form, but it appears to be resistant to bacterial canker.

Populus nigra subspecies *betulifolia*

BLACK POPLAR

Native of east and central England. A large, heavily branched tree belonging to wet valley sites. The trunk is short, with dark, deeply fissured bark and often with large bosses or burrs.

Young shoots are hairy, first green, then ochre-yellow, then dark. Leaves are deep green and smooth on top, pale below. The timber is worthless. The tree is now very rare.

There are about 40 poplar species in the northern hemisphere, but they hybridise very readily and few truly native black poplars may exist. Many poplars suffer from bacterial canker diseases, making them unreliable for woodland planting.

Black Italian poplar (*Populus x canadensis* var. *serotina*) is much more common in Britain than the native black poplar. It is a hybrid between *Populus nigra* and the North American *P. deltoides*, the northern cottonwood. It is a large, vigorous tree, up to 35m, the bole being without burrs. Its leaves, which appear much later than other poplars (*serotina* means 'late' in Latin), are bronze-coloured when they first flush and shiny, later becoming deep, dull green. The buds are large, pointed, reddish-brown and slightly sticky. There are only male trees of this hybrid, which is propagated readily by cuttings.

Black Italian poplar is resistant to bacterial canker and was the principal timber-producing poplar in Britain until the development of the newer hybrids.

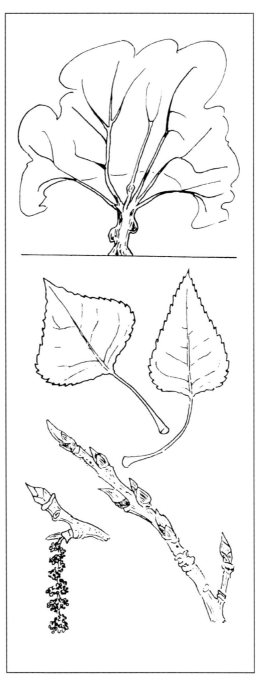

HYBRID POPLARS

Populus species

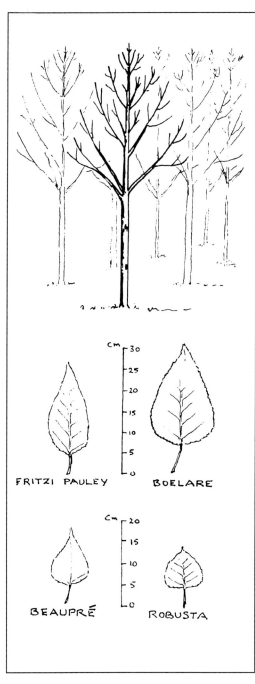

FRITZI PAULEY

BOELARE

BEAUPRÉ

ROBUSTA

Poplar wood is white, tough, lightweight and odourless, and burns without sparking or smouldering. It is used for fruit and other food containers, matches, matchboxes (all by veneering), paper-pulp, wood-wool, truck platforms, carts and other purposes. The main problem in commercial development has been susceptibility to bacterial canker which spoils the shape of the stem and grossly weakens the tree.

Commercial poplar growers look above all for fast, even growth and resistance to bacterial canker, seeking these by selection and crossbreeding. Research scientists have produced new trees with these characters. A plant of proven high quality can be named and registered like a new rose or potato variety. Reproduced vegetatively, the clonal plants are identical, genetically and in growth.

Five groups of poplar clones are currently of special interest, either proved or promising: *P. x canescens*; 'Balsam Spire'; 'Fritzi Pauley' and 'Scott Pauley'; 'Robusta', 'Serotina', etc.; 'Beaupré' and 'Boelare'. (See chapter 14 for details and parentage.)

Since many of these clones have been bred on the Continent, it is not yet clear that they will grow well throughout Britain, especially in the north. The yields of the latest crosses, if successful, are much higher than the unselected parents, so the current trials are important.

Only sheltered moist fertile sites are suitable – good arable farmland. High elevations, exposed sites, shallow and infertile soils, acid peat and stagnant water are all totally unsuitable.

Buxus sempervirens

BOX

Native of southern Europe, the Caucasus and, perhaps, England. (The distribution is curiously discontinuous and the fact that box has been cultivated at least since Roman times makes the native status in England somewhat doubtful.) In England it reaches 9m in height and 20cm in diameter, in the Caucasus 15m and 30cm.

It is evergreen, with small, oval, leathery leaves 2.5cm by 1cm, dark green above and light green below. The flowers are not showy, golden-white and in clusters. The fruits are hard, three-celled capsules, black-brown when ripe, each cell containing two dark brown seeds.

Box sprouts and coppices freely. It is not demanding of nutrition but favours chalk and limy soils which are not too dry. It tolerates heavy shade and is smoke-hardy.

This is a decorative bush and a good hedge plant. The wood is light brown and very stable, not twisting or cracking after it has seasoned. It is a superb turnery wood and in the highest class, with sandalwood and ivory, for carving. Its stability made it valuable for rulers and scales.

Historically twigs of box were used in religious ceremonies, Christian and druidical, which may account for the long and widespread cultivation of the bush.

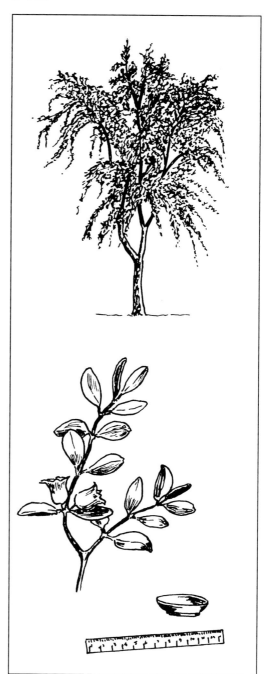

STRAWBERRY TREE   *Arbutus unedo*

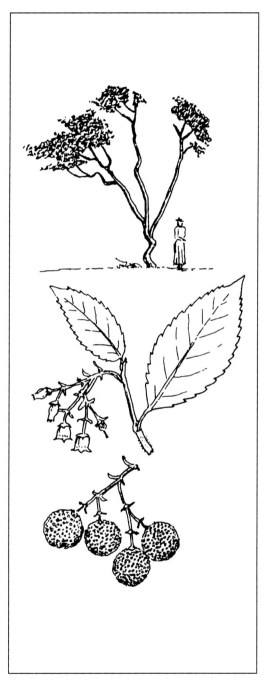

Native in Ireland, where it achieves its best growth (about 12m high), and the Mediterranean, where it is a maquis shrub. It is locally abundant in Co. Kerry as an associate of sessile oak.

The simple, alternate leaves are oval in shape and leathery, dark green above. The flowers are white bells, waxy in texture, and develop into soft red berries, superficially like a strawberry; they take a whole year to ripen and are edible.

A key feature for recognition is the orange-red bark.

The hard reddish-brown wood is used for carving and for the traditional Killarney box-making, but the tree is becoming scarce in Ireland, and is valued for its appearance in parks and gardens in southern England and in Ireland. It is reasonably hardy, although it thrives best in the maritime climate, and is lime-tolerant.

Robinia pseudacacia

LOCUST TREE

Native of eastern USA; introduced to France in 1601 and to England a few years later.

Grows to 25m, with angular branching caused by the failure to develop twig-end buds. The bark is thick, brown and develops broad, deep fissures and ridges in diamond patterns. A relatively short-lived tree.

Leaves are compound, with nine to 21 leaflets, each about 5cm long, blue-green below, oval and smooth. They flush late in spring. At each leafbase, twin spines grow, sturdy and 2–3cm long. The small buds are hidden, entirely enclosed in the leaf-bases; in winter they are sunk in the twig between the paired spines.

The flowers are showy, like sweet peas, sweet-scented, white and hang in long tassels. The fruit is a pod, dark brown, 5–10cm long, containing up to ten seeds. Pods hang on the tree into late winter.

The yellowish timber is tough, resilient and extremely durable. It turns well and makes excellent gate and fence posts.

The tree grows on infertile sandy soils but not on heavy clays or in the wet. The roots have nodules of bacteria which fix atmospheric nitrogen (like many of the pea family). It is intolerant of shade but tolerates moderate atmospheric pollution. It thrives in warmth and in Britain grows well only in the south of England, the south Midlands and East Anglia.

The tree is planted mainly for amenity, as a street and park tree, but is also valuable for planting on mine spoil and industrial wasteland where its nitrogen-fixing capacity gives it an advantage. It coppices well and grows freely from suckers.

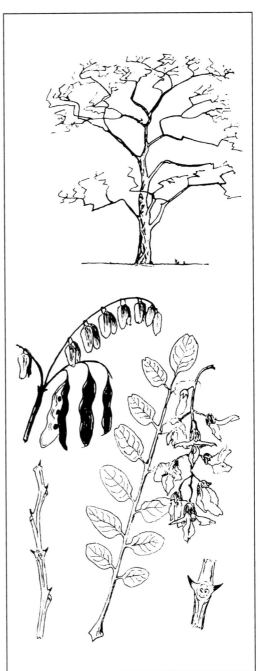

FURTHER READING

Coombes, A.J. (1992), *Trees*, Dorling Kindersley

Dallimore, W. & Jackson, A.B. (ed. S.G. Harrison) (1966), *A Handbook of Coniferae and Ginkgoaceae*, Edward Arnold, London

Mitchell, A.F. (1978), *A Field Guide to the Trees of Britain and Northern Europe*, Collins

Mitchell, A.F. & Jobling, J. (1984), *Decorative Trees for Country, Town and Garden*, Forestry Commission, HMSO

White, J. (1995), *Forest and Woodland Trees in Britain*, Oxford University Press, Oxford

What Trees Can I Grow Here?

4

*Lest the land lie idle, it is useful to plant even in the
mountains.*

Virgil, Georgics ii 37–38

The choice of the kinds of trees to plant on a
given site is probably the single most
important decision for a woodland manager.
If the choice is wrong, no matter how careful
the later management, the result will be
unsatisfactory for the whole of that rotation.
Some species may grow fairly well when they
are young in places where they cannot grow
properly to maturity, so an error of judgement
is not always immediately apparent.

A gardener can make radical changes to the
soil and environment in order to grow plants
with special habitat requirements. In contrast,
the changes a forester can make to a planting
site are quite restricted; breaking a compacted
layer in the soil and improving the local
drainage to increase the rootable depth may
be the limit. Most environment factors must
be accepted more or less as they exist.

Successful woodland management requires
an appreciation of ecology, which in this
context means the relationship of trees to
their environment and to the other organisms
in it. Since trees are long-lived they must cope
not with the *average* weather conditions but
with the likely extremes. The woodland
manager should undertake work to ameliorate
adverse conditions for trees at planting time
by draining surplus surface water, fencing out

browsing animals, cultivating, providing
fertilisers to relieve immediate nutrient
deficiencies and others. Nevertheless, these
will not allow the manager to bypass the
ecological constraints which dictate which
species can succeed.

On fertile land and in a kind climate, a
wide range of trees, perhaps 30 or 40 species,
will grow successfully. Progressively, as one
moves to less fertile soil and harsher climate,
the choice narrows. On an inhospitable site
it may be that only two undemanding
species may be planted with any chance of
success. On land poorer still, although the
same two may be established by careful and
expensive work, even they will not grow well
– and certainly not well enough to give a
positive financial return on the investment
involved in planting them, although there
may be other benefits, such as shelter.

Trees which are demanding should not be
planted on infertile soils. Although those
which are undemanding in respect of fertility
can be planted and may grow fast on fertile
soils, it is generally unwise to plant them
there, because of both the wasted
opportunity and, often, the poor form of the
grown tree. For instance, when planted on an
unduly fertile site Japanese larch grows stems

with a corkscrew shape which are disastrous for marketing, and lodgepole pine produces exceptionally weak wood and coarse branches; they are better on less fertile soils.

Excellent guidance on the choice of trees to plant on different sites throughout the British Isles was published some 40 years ago by Professor M.L. Anderson in a small book, *The Selection of Tree Species*, which still offers good advice to the planter of smaller woodlands. Anderson's system was based on two easily understood characters of bare sites where trees might be planted: wetness and fertility. In respect of moisture, the soil is described as dry, moist, wet and wet-with-peat. Soil fertility may be recognised by common plants growing there, which are indicators. On the basis of this guidance and experience, the choice of species is shown in the table on pages 110 and 111.

In choosing trees to plant in a particular place, an important consideration is what the planting is for. The objective may be shelter against the wind, or to screen a development, or to provide firewood for home use or as timber to sell, or for landscape amenity, or as game and wildlife habitat. It may be possible to achieve several of these purposes but you should begin by being clear in your own mind about what the main intention is of planting trees in a particular place. It is the first question a good consultant would ask you, so it is the first question you should ask yourself. In effect, the objectives must be considered against the set of trees which are ecologically feasible for the site.

The growth and vigour of mature trees growing on truly comparable sites nearby provide guides which should not be missed, although they may have been planted with quite a different objective from the current ones. They are evidence of what can succeed and what does not grow well, but do not themselves close the case.

It is a common misconception that broad-leaved trees always grow more slowly than conifers; in fact some grow in height very rapidly on sites which suit them. Admittedly, conifer thinnings are usually easier to sell than broadleaves, their rotations for timber production are usually shorter (the period from planting to maturity) and the marketable volume of their timber production is usually greater. Nevertheless, it is seldom sensible to have simple volume production as the main objective in managing a small woodland.

Can the grower hedge his bets by planting more than one species? The main advantage of a pure crop is its simplicity. It is easier and cheaper to plant one species than a mixture, and generally easier to manage and simpler to market at maturity, provided that the choice was a good one. Nevertheless, pure tree crops score low points in naturalness and possibly in ecological efficiency. It is unusual to find only one tree species forming a natural forest over more than a small area; others usually find a niche which they can occupy and an argument strongly pressed in favour of mixed planting is that the site is more efficiently used if several species are present. Different soil drainage, fertility and aspect can be exploited to accommodate different trees, and it may well be true that more than one species will be more efficient in intercepting the sunlight for its conversion into wood, for example oak with a beech understorey.

Mixtures may be planted for aesthetic reasons, for instance a proportion of larch planted with an otherwise pure crop of Sitka spruce, in order to provide a difference of colour and texture in a commercial planting. This mixture also benefits tree nutrition, the nitrogen status of the spruce being markedly better than in a pure crop, especially on infertile heathery sites.

In some places and circumstances,

mixtures may be planted so that one species, usually a pioneer such as birch or Scots pine, assists another more valuable one, the former acting as a 'nurse'. The nurse species may be more frost-resistant and able to protect the more tender one in early life, or it may provide side shelter and enable a more valuable tree to grow faster or with better form. Birch, alder, aspen, Scots pine and larch are often used as nurses for oak, beech and Douglas fir. Many of the best mature broadleaved stands we admire today were planted as mixtures, such as oak with pine or beech with larch, and one of the great advantages offered is the early cash income which the owner may receive from cutting and selling the conifers. Planting of broadleaves with Norway spruce, which are cut as Christmas trees in about year seven, is the extreme example of planning an early return from a mixture.

Mixed plantings create difficulties when there is a marked difference in the rate of growth between the species. The more intimate the mixture, the greater the risk of the nurse suppressing the main crop, and weakness or delay in tending work may result in complete frustration of the aim to nurse the potentially more valuable trees. Small groups of one species in a matrix of another, and planting alternate strips of two species (three rows of A, three of B, three of A, etc.) are preferable to intimate mixtures of species which may result in the complete swamping of the more slow-growing. Strip mixtures with a minimum of three rows of conifer alternating with three or more rows of broadleaves appear fine on paper (although unattractive when seen across a valley later) but require very prompt action when tending is required, specifically when the nurse begins to compete with the main species instead of assisting it. If the difference in early growth rates is such that

rows of the nurse have to be removed before they are marketable, the cost of the necessary tending is very high.

For most small woodland planting, group mixtures probably serve best. They are sound ecologically and are more acceptable visually than lines, both within the wood and from a distance. The size of group can be varied according to species and site: the hardier and more light-demanding the species, the larger the groups may be; the more tender to frost or exposure and more shade-tolerant, the smaller the groups may be. It is an advantage of a group structure that each group forms a clear unit of work and decision for subsequent tending. In order to accommodate variations of site conditions and for reasons of visual amenity the pattern of groups over the site may be irregular, but within each group a regular planting pattern greatly simplifies the task of finding the plants among weeds. For ease of tending, the centre of each group should be marked with a recognisable stake, retained until the plants are established. The smallest group should usually be nine or 16 plants, the groups being at about 10m or 12m centres. A group of a light-demanding species, such as ash, might be up to 49 plants, each group to provide two or three final-crop trees.

Firmly avoid the practice adopted in some motorway tree-planting where eight or ten species are jumbled together in intimate mixture; it is wasteful and a nightmare for tending. It is a mistake even if the objective is purely for amenity, to mimic a wild wood of native species; in a natural forest it is usual for trees of one species to arise and grow in groups, as seedlings find a patch of ground suitable for germination after a heavy seeding.

In devising the planting pattern of lines or groups, try to envisage what conditions will be like when the nurse trees are cut and what, if anything, may be done in the spaces created. It may be appropriate to use the space

SITE

SOIL	TYPICAL PLANTS
BROWN EARTHS Fertile valley bottoms and lower hill slopes	Dog's mercury, wood sanicle, broad buckler fern, male fern and lady fern
ACID BROWN EARTHS Well-drained soils, with moder humus	Fine heath grasses, sheep's fescue, wavy hair grass, sweet vernal
CALCAREOUS SOILS Chalk and dry limestone areas	Lime lovers: thyme, calamint, false brome grass
LOWLAND GLEYS Heavy fertile silts, limestone clays (mainly southern)	Hard rush (*Juncus inflexus*), sedges (*Carex spp.*) and jointed rush (*J. articulatus*)
PODZOLS Glacial sands and gravels, including ironpan soils after ripping to break the pan	Purple bell heather (*Erica cinerea*), heather (*Calluna*), fine grasses, dry mosses, vaccinium communities
UPLAND GLEYS without peat Clays, moderately fertile, tending to be acid	Jointed rush, matgrass (*Nardus stricta*)
FERTILE PEATS Fen and carr peats, and groundwater gleys	Willow-reed community, alder carr with tussock molinia, soft rush (*Juncus effusus*)
FLUSHED PEATS and **PEATY GLEYS** Peats flushed with drainage water from higher land with nutrients	Molinia peat, cottongrass peats, *Eriophorum spp.* with *Juncus squarrosus*
UNFLUSHED PEATS Flat and basin bogs, deep blanket peat	Cross-leaved heath, (*Erica tetralix*), deer-grass (*Trichophorum caespitosum*), sphagnum mosses

TREES TO PLANT

BROADLEAVES

Ash, oaks, beech, walnut, lime, gean, sweet chestnut, sycamore, Norway and field maples, grey poplar, aspen, hybrid poplars (in south), hornbeam, whitebeam
Understorey: hazel, holly, bird cherry, wild apple
Nurses: common and cordate alders, birch

Sessile oak, beech, birch, gean, sweet chestnut, sycamore, holm oak
Understorey: hazel, holly, bird cherry

Beech, sycamore, ash, lime, grey poplar, aspen, grey and cordate alder, Turkey oak, hornbeam, whitebeam, horse chestnut, gean, Norway maple, field maple, wild service, box

Ash, pedunculate oak (south), sessile oak (north), hybrid and grey poplars, sycamore, gean, crack and white willows, horse chestnut, London plane, hornbeam, Turkey oak
Understorey: hazel, Swedish whitebeam
Nurses: common and cordate alders, aspen, birch

Red oak, silver and downy birch, rowan, Swedish whitebeam

Sessile oak, aspen, downy birch, goat willow, rowan, Swedish whitebeam
Nurse: alders

Hybrid poplars (south only), alders, bay and white willows, ash

Black alder, goat willow, downy birch, rowan

Downy birch

CONIFERS

Douglas fir, grand fir (sheltered), Lawson cypress, spruces, western red cedar, larches, western hemlock, noble fir, wellingtonia
Understorey: yew
Nurse: Scots pine

Douglas fir, larches, spruces, Scots pine (in north), Corsican pine (in south), grand and noble firs, juniper, western hemlock, wellingtonia

Corsican pine, Norway spruce, Japanese larch, western red cedar, Atlas cedar (where sheltered)
Understorey: yew, juniper

Scots pine, Corsican pine, Norway spruce, western red cedar, noble fir, western hemlock, Lawson cypress

Scots pine, hybrid larch, western hemlock, Corsican pine (lowlands)
Understorey: juniper

Spruces (Norway, if non-peaty), western hemlock, noble fir
Nurse: Scots and lodgepole pine

Norway spruce

Sitka spruce, lodgepole pine

Lodgepole pine

The simplest pattern for group planting comprises groups of 9 trees set in a matrix of another species. If there are 3 lines of matrix trees between the groups and plants are at 2m x 2m, there are 72 groups per hectare. With only 2 rows between the 9s, also at 2m, there are 100 groups per ha. Spacing of groups depends on species and size at maturity.

Or groups may be larger, say 21s. It is easier if they are kept symmetrical.

If regularity of the group pattern is visually undesirable, they may be randomised. Here are 3 species in equal amounts and random groups of 9 trees.

GROUP PLANTING PATTERNS

for planting a shade-tolerant understorey such as hazel. If the intended final-crop trees are being planted in groups, the spacings between groups must match the size of the crowns of that species at maturity (e.g. for large oak, sycamore, etc., about 12m in diameter). A small group will provide only one final tree; larger groups must be of a size to provide several. If in doubt look for a simple design which will be robust in practice; an over-elaborate design may be self-defeating.

Some planting may be done where exceptional conditions must modify the normal advice, and the diagnosis of Table 4.1 is incomplete. Some special conditions are considered below.

ATMOSPHERIC POLLUTION
Industrial fumes and smoke damage trees. Point sources of pollution are nowadays much

less common than in the past, when trees were killed in the immediate vicinity of metal smelters and coke works. Nevertheless, atmospheric pollution in and around cities and alongside motorways constrains the selection of trees for successful planting. Sooty deposits clog stomata on the leaves and reduce the reception of light for photosynthesis. Oxides of nitrogen can prevent the stomata of some species operating properly, in effect jamming them open, so that the tree cannot control its loss of water, resulting in a severe loss of foliage, as in the thin crowns of spruce in cities.

Where atmospheric pollution is severe, deciduous trees are more tolerant than evergreens; they grow a fresh set of leaves each spring. Do not plant spruces, silver firs, Scots pine or the larches. Reliable trees are listed in Table 4.2.

Table 4.2

TREES TOLERANT OF LOW AIR QUALITY

Field maple	Cordate alder
Norway maple	Common alder
Sycamore	Grey alder
Holly	Crab apple
Grey poplar	Red oak
Whitebeam	Turkey oak
Swedish whitebeam	London plane
White willow (incl. Tristis)	Common lime
Horse chestnut	Large-leaved lime
Aspen	Lawson cypress

FAIR BUT LESS RESISTANT

Red alder	Silver birch
White poplar	Gean
Rowan	Bird cherry
Atlas cedar	Corsican pine
Hawthorn	Walnut
Ash	Red chestnut
Beech	Locust tree

SALT SPRAY

Table 4.3 lists species suitable for planting in places subject to salt spray, which may be seaside areas or strips near main roads on which salt is spread in winter frosts. Norway spruce is very susceptible to salt damage, Sitka less so.

Table 4.3

TREES TOLERANT OF SALT SPRAY

Sycamore	Sea buckthorn
Hawthorn	Midland thorn
Holly	London plane
White poplar	Black poplar
Grey poplar	Whitebeam
Turkey oak	Evergreen oak
Corsican pine	*Pinus radiata*
Goat willow	(south-west only)
Lawson cypress	White willow

LESS TOLERANT TREES

Gean	Limes
Atlas cedar	Swedish whitebeam

SEVERE WIND EXPOSURE

Recommendations for the design of shelterbelts are given in chapter 12. Table 4.4 lists species which are able to tolerate exposure to wind, without considering their place in shelter-planting.

Table 4.4

TREES TOLERANT OF WIND EXPOSURE

Sycamore	Field maple
Birches	Ash
Lodgepole pine	Sitka spruce
Wellingtonia	Noble fir
White poplar	Sessile oak
Grey poplar	Turkey oak

Swedish whitebeam	Evergreen oak
Lawson cypress	Whitebeam
Leyland cypress	Rowan
Hawthorn and	Hybrid larch
Midland thorn	Yew
Holly	

Beech grows in some extremely exposed places but is not listed in Table 4.4 because it can become exceedingly distorted by wind. When planted on the lee side of a belt, the branches of its flag-crowns commonly sweep over the neighbouring field, causing much loss of production there. If it is planted to give shelter or other benefits on farms, it should be restricted to places where its crown will not overhang fields.

INDUSTRIAL WASTELAND

Any substantial tree-planting on these sites should be guided by an experienced professional forester. Even for small-scale work it is particularly difficult in a book to recommend species to plant on industrial wasteland, since the soil may be highly acid or highly alkaline, and it may be very infertile, compacted or toxic with heavy metals. Trees to cope with such conditions are mostly natural pioneer species. Shortage of available nitrogen is a common problem on slag-heaps; the alders and the locust tree have the capacity, with associated micro-organisms, to fix atmospheric nitrogen, thus overcoming the problem. Colliery waste is usually acid and may require the application of up to 40 tonnes of lime per ha. If you are in doubt about the nature of the waste (its acidity or the presence of toxins), get the soil analysed before work begins. All the trees in Table 4.5 will not grow on all kinds of industrial wasteland, but they are candidates. Do not expect any trees to thrive on sterile or toxic soil; at best they must struggle to establish the beginnings of a woodland ecosystem, so that

their own leaf-litter may be recycled. Anything which assists that should be encouraged. For instance, the importance can scarcely be overstated of mycorrhizas, the beneficial soil-living fungi which live in close association with the fine roots of trees; they are part of the woodland ecosystem's recycling process and without them many trees can barely function. Good results in inoculating wasteland with the necessary fungi have been achieved by importing small amounts of litter from a healthy woodland to spread around the newly planted trees.

Table 4.5

CANDIDATE TREES FOR USE ON INDUSTRIAL DERELICT SITES

Sycamore	Ac + Al + D + M
Grey alder	Ac + Al + D + M
Silver birch	Ac + Al + D
Red oak	Ac + D
Rowan	Ac + Al + D + M
Corsican pine	Ac + Al + D + M
Hawthorn	Ac + Al + M
Common alder	Ac + Al + M
Red alder	Ac + M
Downy birch	Ac + M
Goat willow	Ac + Al + D + M
Lodgepole pine	Ac + M
Scots pine	Ac + Al + D + M
Locust tree	Ac + Al + D
M = moist	Ac = acid
D = dry	Al = alkaline

CHOICE OF SPECIES

When you have a list of possible species to plant (you can use chapter three to check the preferences and requirements of individuals), the final choice depends on your objectives and on any constraints placed by nature conservation agencies (such as areas designated as Sites of Special Scientific Interest).

On easy sites – fertile, moist, sheltered – it is perfectly sound to plant the trees which will form the final stand, although perhaps with the addition of some to give an early income. As difficulties increase, the use of nurse trees becomes good practice, in order to help less robust species through the early years. On really difficult sites it is often wiser to be patient and to plant a pioneer species; in the lowlands this might be an alder, a tough poplar such as *Populus alba* var. Racket (not a high-productivity clone) or a willow such as goat willow or *Salix x dasyclados*; in the uplands downy birch, Scots pine, lodgepole pine or Sitka spruce might be used for the same purpose. When the pioneer has established woodland conditions over the area, groups may then be cut out and planted with the species wanted for the long term. This strategy may not take all that long and may give a much better final result than any other. It has been used very effectively in Belgium and the Netherlands, on land reclaimed both from industry and from the sea, although less commonly in Britain. The use of a quick-growing pioneer crop is to be particularly recommended where people are creating their own community woodland; the successful growth keeps the group enthusiastic through the establishment phase and they can discuss the design of the wood they really want far better than on a bare site. On really difficult sites one cannot afford to be idealistic and try to get only trees of high value from the start. If the plant is upright and woody, accept it; the stand can be enriched later.

The species chosen for planting must depend on the soil and other factors on the particular site, but the following points may be helpful:

- For timber production, small woods must be managed for quality of product and specialist markets, not for low-quality bulk, because in a small woodland the 'bulk' is apt to be unattractively small (see chapter 17). Continually managing for the highest quality the trees can produce will give future managers the widest options. At a pinch the grower can burn veneer butts to keep warm, but is unlikely to make Rolls-Royce facias out of firewood.

- Consider how work might be scheduled and the wood structured to give interim returns early in the rotation.

- You *could* plant a large area uniformly. That is simple, but it means the whole area will come up for every tending at the same time, and, if hardwood timber is aimed at, there may be a long wait for the first income.

- In all but the smallest patch of woodland, consider a group structure, even planting sections of the wood over several years. Groups planted at the start could include some long-rotation species – say oak – if that is the intention, some shorter rotation – perhaps ash and sycamore, some short rotation – say hybrid larch, and some ultra short-term for Christmas trees and foliage – say Caucasian fir or noble fir. That would be much less straightforward, but it would offer two advantages: tending and harvest operations come in group-sized parcels (a nightmare for the manager of a large forest but a blessing in small woodlands worked with home labour); and it offers a sequence of incomes from an early date – Christmas trees, greenery, small stakes, fence posts and so on – while high-quality hardwoods are growing and promising for the future. In a later chapter I shall write about the farmer John Dixon and you will see this is not a fancy theory but solid practice that has worked well. In creating his 10ha wood, John planted as little as 0.4ha in some years, the area he could handle. It is a strategy which fits excellently with multi-purpose woodland and can be designed mainly for broadleaves or conifers, depending on the locality.

- The more fertile and more sheltered the locality, the greater the opportunity for early income – but also the greater the penalty may be for not attending to maintenance and thinning when they are really required.

CHOICES FOR WOOD PRODUCTION

If you intend to plant in an ancient woodland, you may want to restrict yourself to species which are native, at least for the main stand, although there may be a case for non-native nurses if the woodland already contains exotics. The generally accepted native trees are listed at the beginning of chapter three, but you should remember that all are not native to all parts of Britain. Should you require guidance about the local list and sources of seeds or plants which are truly of local origin, the Forestry Authority and the country's nature conservation agency will help.

A rigid restriction to native trees outside ancient woodlands and conservation areas would impose a severe financial penalty, because many native species are not high producers of timber volume and some of their wood requires highly innovative marketing if it is to sell for a good price. More use should be made of birch, alder and others, provided that trees of really good form can be grown, so that pricey markets develop. Well-grown ash, yew, beech and oak, once they are of marketable size, sell well and are likely to do better in the future but the penalty for the grower arises in the period of waiting, which may be long, until the trees reach sufficient size to be in demand. A sensible balance is required: the past relative neglect of native trees was unwise; a doctrinaire rejection of all non-native species outside designated conservation areas is equally so.

Inevitably, in the real world, the choice of silvicultural practice is deeply influenced by finance. For instance, there is a strong desire among people to see more broadleaved woodland, especially oak woods grown on a long rotation, and there is a wide expectation that these should be planted as pure oak, unadulterated, as it were, by other species. Usually, however, young thinnings of oak are difficult to sell at a profit. Financially, planting pure oak may be compared to putting money, say £4,000 for each hectare, into a bank account which not only pays no interest but requires the holder to make further pay-ins for maintenance, perhaps for 50 years. In fact, the account may not produce a positive return until the trees are mature at 150 years. With that prospect, the bank manager (and certainly the Treasury in respect of national forests) would say that, as an alternative to pure oak, the £4,000 put into a savings account for 150 years with interest would grow to £2 million, whereas the hectare of oak timber would be worth only about £100,000.

Even with planting grants available at present rates and allowing for the fact that timber income is not the sole benefit, the only way, realistically, in which high-value hardwoods can be grown, so that future generations may enjoy them, is for growers to ensure that the woodland yields early incomes. For broadleaved woodland conservation it could be financially disastrous and self-defeating for foresters to be denied the use of species capable of providing effective intermediate money yields. It is not enough to receive a grant which pays for the planting; the stands must be tended, decade after decade.

The forester aiming to produce large broadleaves in small woods in the lowlands should try to plan for a sequence of yields, some of which must come, almost inevitably, from conifer species and, later, from the faster-growing high-value hardwoods such as ash and sycamore. By such a strategy the long-production-period hardwoods can be restored. It may be the only way in which they can be. Many of the much-admired present stands of old broadleaved trees were grown with conifer nurses in their youth, so the system works well. Above all, the woodland needs to benefit from the attention of an interested and caring owner (which is as true for the state forest as for the small, private woodland).

CALCULATION OF PLANT REQUIREMENTS

The number of plants required for each project must be carefully calculated. The number of groups multiplied by the number in each group will generally be the basis of the calculation, but where there is an area of straight planting and as a check of the overall total, the figures in Table 4.6 may assist. In the table the number per hectare is obtained by dividing 10,000 by the square of the distance in metres between plants. Since one cannot plant right up to the fence lines or in the drainage ditches, a smaller number of plants than this will be needed in practice. Experience shows that these areas and the provision of roads and loading spaces on average amount to 15 per cent of the gross area of the woodland. The planting grant is paid in the full area and allows a proportion of ground to be left unplanted for wildlife conservation and archaeology.

Table 4.6

NUMBER OF TREES PER HECTARE AT COMMON SPACINGS

Distance apart in metres	Trees per ha.
1.5 x 1.5	4444
2 x 2	2500
2.2 x 2.2	2066
2.4 x 2.4	1736
2.7 x 2.7	1372
3 x 3	1111

A guide to the sizes of different classes of plant and the relative costs of a typical broadleaved species are shown in Table 4.7. The price differentials are expressed as indices, based on 1997 ex-nursery lists and on the price of a 1+0 seedling (=1).

Table 4.7

TYPICAL SIZE CLASSES AND PRICE DIFFERENTIALS OF TREES

Plant type	Size	Prices ex-nursery
1 + 0 seedling	10–15cm	1
Transplant 1 + 1	20–30cm	1.3
Transplant 1 + 2	30–40cm	1.7
Whip 1 + 2	60–80cm	3
Whip 1 + 2	80–100cm	5
Feathered whip	150–180cm	7
Feathered whip	180–210cm	15
Half standard	210–240cm	20
Light standard	250–275cm	32
Standard	275–325cm	44
Heavy standard	4 metres +	250–6,000
Cell-grown plant	25–45cm	2
Cell-grown plant	45–60cm	3

Over-specification of contracts by landscape advisers, especially of public works, is a problem, partly because of the misconception that big plants are best. Table 4.7 shows the effect of specifying, say, feathered whips where transplants would be satisfactory (or better); since the adviser's fee is probably 17.5 per cent of the contract price, one can easily see why some advisers recommend big plants. Occasionally it is useful to plant large trees, but seldom is the price justified by the result.

The use of tall plants is particularly common in urban tree schemes; unfortunately their size makes newly planted trees very obvious and attractive to vandals. Small transplants are less obtrusive and less satisfying for vandalism; their lower cost makes it possible to plant many more trees and replacement of failures is much easier. The use of small plants makes possible 'tree cover by stealth' in peri-urban areas, comfortably accepted by people because the change in their environment is gradual.

The establishment of the wood must be seen as a whole: the preparation of the site, the selection of the species, the type and most appropriate size of plants, the tending, protection and perhaps the fertilising of the stand. If one element of this sequence is skimped, a higher price must be paid at the next stage in order to retrieve the effective establishment. And if the establishment is poor or the wrong species is planted, it will take a long time to recover the position.

FURTHER READING

Anderson, M.L. (1950), *The Selection of Tree Species*, Oliver and Boyd, Edinburgh

Dobson, M. and Moffat, A. (1995), *Site Capability Assessment for Woodland Creation on Landfill*, Research Information Note 263, Forestry Commission, Edinburgh

Evans, J. (1984), *Silviculture of Broadleaved Woodland*, Forestry Commission, Edinburgh

Moffat, A. and McNeill, J. (1994), *Reclaiming Disturbed Land for Forestry*, Forestry Commission, Edinburgh

When ye hae naething else to do, ye may aye be sticking in a tree; it will be growing, Jock, while ye're sleeping.

Sir Walter Scott

The Laird of Dumbiedikes in *Heart of Midlothian*, 1818

In spite of the fact that the planter must pay high regard to the 'unchanging' features of the planting site, there are actions which may marginally improve the habitat conditions and may thus both widen the choice of species and help the trees' establishment, such as draining, cultivating and fertilising.

DRAINING

Tree roots cannot survive without oxygen and the rootable soil is restricted by the water table which excludes the air. A site which has a water table close to the surface or which is liable to periodic flooding will always be difficult for tree-growing and its productivity of wood will be low. In order to improve these conditions the water table in the soil may be lowered by draining or conditions for young trees helped locally by mounding.

Drains in woodland should always be open ditches. Agricultural tile drains or perforated pipe drains buried in the soil soon become blocked by tree roots growing there because of the good aeration; consequently pipe drains must be avoided. Once well established, a fast-growing tree stand will lower the water table by transpiration and by the canopy intercepting rain which then evaporates from the foliage.

Water runs directly downhill and percolates through the soil in the same direction, so ditches should run as far as possible to intercept these flows, that is to say, along the slope with a slight fall to carry the water away. The slope of the ditch bottom should not exceed 2°, or 2cm in a metre; with a fall greater than this there is a danger that the ditch will become a torrent in wet weather and will erode into a gully. A fall of 2° is sufficient to keep the drain reasonably clear of standing water and of accumulating debris and weeds.

Percolation through fine-textured soils, such as silts and clays, is very slow and generally restricted to the upper layers. On these soils there is little point in digging drains at close spacing, since the water is firmly held in the soil structure. The drains should be planned to cut off the supply in the more permeable upper layers and also to remove it from hollows. It may be that no drainage will be required in sandy soils, chalk or limestone.

The sides of open drains should have a batter (i.e. a slope outwards) of about 60°, although in stiff clays the sides will stand at 70°. The minimum width for the bottom of the drain will be the width of a spade or

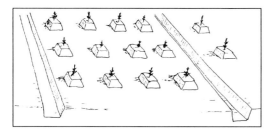

1930s UPTURNED-TURF PLANTING
The turves were spread by hand from the hand-cut drains. Tens of thousands of hectares of land were prepared for afforestation in this highly labour-intensive way between 1925 and 1950.

shovel, 20–25 cm. Drains may be cut by hand or by excavator.

The small woodland planter should not attempt to plant trees on deep-basin peats or raised bogs. For draining shallow peat and blanket bogs there are special hand tools, notably the *rutting spade* and the *hack*, the former to slice or saw through the peat to cut the sides of the drain and to cut the ribbon between them into sections, and the latter to drag out the sections. This hand-cutting is very heavy work indeed.

CULTIVATING

For nearly the last 50 years large-scale afforestation in Britain has been founded on the use of ploughs and powerful tractors to pull them. Ploughing has allowed the effective removal of water, the preparation of planting positions for the individual trees and, on hard ground with compacted or ironpan soils, cultivation of the soil down to about 60cm with a tine or ripper.

The use of ploughs, often with an application of phosphate fertiliser to the soil, has had the effect of evening out the differences between sites. Before ploughing was in general use the trees to be planted had to match the natural habitat conditions and a wide diversity of species resulted. With

ploughing for upland afforestation it has been possible to plant fewer species, even only one over large areas. For good economic reasons Sitka spruce is favoured (especially for the uniformity of the product and the high demand for its timber) and the result is near-monoculture over quite large areas of the uplands.

Useful though it has been in large-scale afforestation, ploughing has also been found to have some severe disadvantages. In high-rainfall areas the up-and-down hill furrows of tine ploughing have sometimes become erosion gulleys, although generally successful on other sites. Further, since tree roots do not cross furrows, the special ploughing to prepare planting positions for trees on peaty sites so strongly encourages the trees' root systems to run along the ridges of plough spoil that the trees have inadequate anchorage to resist strong winds at right angles to the furrows

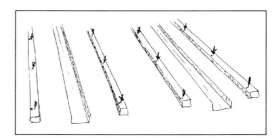

DOUBLE MOULD-BOARD PLOUGHING
A very common preparation for planting upland peats between 1955 and 1985. The soil from the plough furrow was cut in two and thrown by two mould-boards, one on either side of the furrow. The furrow was normally not a drain but merely a source of the upturned turf.

and the stands are basically unstable. The 'domino effect' can lead to severe windblow.

As the value of ploughing as a preparation for tree-planting has been questioned, attention has turned to the use of scarifiers,

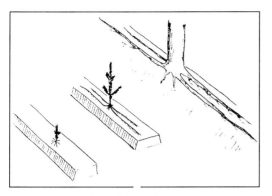

PLOUGH RIDGE ROOTING

*The architecture of tree roots is set very early in life.
On wet ploughed land the fine roots inevitably grow
along the upturned turf ridges which are drier, and
that is the pattern for the tree's life. In severe winds, a
tree with such a linear root system is much less stable
than one with all-round rooting.*

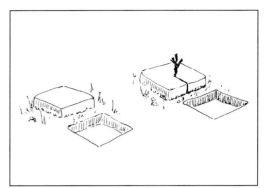

UPTURNED-TURF PLANTING

*A turf or a mound can be cut locally (instead of from
drain spoil). The tree may be planted in the turf if it is
thick enough, or the roots may be spread in the
sandwich below the turf by cutting a slit from one side
to the centre.*

mounders and excavators as methods of cultivating the planting site for trees, to break up compacted soil and increase the rootable volume. In the 1930s, before forestry ploughing was general, huge afforestation schemes were achieved on peat by planting on slices of turf cut from the drains, so-called upturned turf planting. This is still good practice, although it is hard work. It is also the basis for the use of long-reach excavators which can cut a trench and place each bucketful of spoil as a mound at the place where a tree is to be planted. The trench may be a continuous one to form a drain or it may be a short one purely to provide soil for planting positions. The bucket excavator making planting mounds is well-suited to preparing a felled woodland for replanting; the machine can push debris into the trenches in the same operation.

The upturned turf and the mound from an excavator or scarifier are platforms of soil which, because they are raised above the surrounding ground level, drain well and provide a fair rooting medium for newly planted trees, allowing all-round rooting and avoiding a major disadvantage of ploughing. If trees can establish their roots and begin shoot growth, they are usually well on the way to dominating the site and their roots will invade the undisturbed soil from their initial bridgehead. The good start is essential. The upturned turf also offers the advantage of a short period free from close weed competition.

If the woodland to be planted is quite large it may be feasible to hire an excavator and driver to do this work, but on a small scale a similar result can be achieved by hand using a mattock. The scarifier is a power-driven implement designed to break the ground surface and prepare individual planting sites without forming continuous ridges or ditches. Agricultural subsoilers can serve a similar purpose.

HERBICIDE USE

Competition from weed growth, especially grass, is a major factor in poor survival and poor growth. A turf of grass stops almost any

EXCAVATOR MAKING MOUNDS FOR
PLANTING
*Mounds can be made with a spade or mattock. Larger
plantings require an excavator.*

water reaching the roots of the young trees
and they will grow scarcely at all over many
years. For success they must be relieved of this
close competition. There are various ways to
prepare the site so that planting is done later
in weed-free patches. A pre-planting herbicide
may be used, usually in the summer before;
alternatively a grass sward may be screefed (i.e.
a thin layer of turf is sliced off), although this
is seldom as effective.

The application of a pre-planting herbicide
in 1m² patches at each planting position or in
strips is the most effective treatment.
Certificates of competence are required by
contractors, including volunteer groups, using
all herbicides except one or two expensive
ones for garden use, unless they are working
under the personal supervision of a
certificated person. The Agricultural and the
Forestry and Arboriculture Training Boards
provide information on courses.

In preparation for planting, grass swards
and similar sites may be treated with a
commercial herbicide. Some, such as Atlas
Lignum, must be applied in spring in
preparation for autumn planting; others, such
as glyphosate (marketed as Roundup), may be
applied much closer to planting time, or to

treat regrowth. Glyphosate is effective against
almost all plants and, if it is used post-
planting, the trees must be protected from the
spray because it will damage or kill them. (On
contact with the soil, Roundup is deactivated
and quickly broken down, so that planting
can be done soon after a preparatory
application and there is no lasting
environmental effect.)

Apart from grass, brambles and
rhododendron probably pose the most severe
weed problems and must be tackled before
planting. Brambles may be crushed, torn out
and cut, but heavy growth is most effectively
treated before planting with a herbicide, of
which the most effective at present is probably
imazapyr (Arsenal 50F). If rhododendron is
more than scattered seedlings (which should
be ripped out), the bushes should be cut and
all stumps and shoots thoroughly treated with
the herbicide. Regrowth must be treated
rigorously. Rhododendron has a toxic effect
on the soil around it and where it is rampant
(it may form a solid jungle 5m high on some
sites) it would be best to seek advice on the
most appropriate current treatment from the
Forestry Authority or from a registered
forestry consultant.

Bracken competes with young trees for
light during the summer and in the late
autumn it collapses and tends to smother
them. It is best treated before planting, on a
small scale by breaking and on a large scale by
herbicide. Just as the fronds are unfolding the
stems are brittle and easily broken with a stick
or a length of wire used as a flail. The most
effective herbicide presently available is
asulam (marketed as Asulox), which is taken
up by the foliage and translocated to the
roots. In the year it is applied bracken may
appear to be little affected but it is very weak
or fails entirely the following year. Trees
should not be planted for at least six weeks
after applying asulam. It is essential to get the

trees growing strongly after planting so that they quickly dominate the site, as the herbicide controls the bracken for only two to four years. If asulam is to be applied after trees have been planted they must be protected; although most conifers are tolerant, most broadleaves are not.

In using any chemical as a herbicide it is essential to follow exactly the instructions on the container and to take proper safety precautions to protect yourself, other people, animals and the environment, not least in disposing of containers and washing equipment. The Forestry Commission has a specialist publication entitled *The Use of Herbicides in Forestry*, Field Book 8, which it continually updates as products and practices change. A chartered forester should be approached for up-to-date advice before anyone uses herbicides for the first time in woodlands. Take note particularly that herbicides approved for agricultural and horticultural use are not all approved for use in woodlands (or vice versa). Take qualified advice and follow the rules.

SPACING

Since 1950 there has been a trend to increase spacing at planting from 1.4m to 2.5m and more. The decision by the Forestry Commission that planting grants would not be paid for planting at more than 3m (a standard now changed) very regrettably resulted in broadleaved trees being planted at 3m as the norm instead of the maximum. The advantages of wide planting were seen as cheapness and delay in the need for thinning.

These were undoubtedly attractions for 'big-time' forestry but they should not be for the small grower. Value for money is more important than simple low cost; early yields from thinnings are usually an advantage to the small grower and part of his value-for-money calculation, in contrast to the attitude of the large grower for whom early thinning is a costly embarrassment, yielding too small a timber volume to be financially attractive in extensively managed plantations with contract labour.

Wide spacing means that virtually every plant must grow, otherwise the gaps are seriously large. Canopy closure is delayed so that it takes longer for the trees to dominate the ground plants and the weeding period is longer. Wide spacing allows branches to grow longer and heavier before they are constrained by neighbouring trees, so knots are larger and the timber quality is lower; it also means that the rings of early timber growth are wider and the proportion of weak juvenile wood in conifers is greater. And there are fewer trees from which to select the final mature stand. In most woodland planting (excluding hybrid poplars and the like), wide spacing is initially cheap and later leads to a low-quality result.

Plant spacing should not be wider than 2m where the grower is aiming at quality timber production. This applies even to the strong-growing Sitka spruce if it is intended for construction timber. Slower-growing conifers, including Scots pine, may be planted at 1.8m, and planting at that spacing or closer is advisable on difficult sites, when interplanting different species to allow very early harvesting of a special crop, or when planting oak pure. Broadleaves are normally planted closer than conifers, 1.4–1.8m being preferred, although gean may be planted at 2m by 2m or even wider, and hybrid poplars wider still (see chapter 14). Spacing can be varied with species within one planting area. A group of oak might be planted at 1.4m spacing, while a species planted around the group for nursing, shelter and diversity might be wider. A regular pattern of planting is desirable because it helps in finding the small plants when weeding.

DIRECT SOWING OR PLANTING?

In almost all circumstances planting is more successful and more cost-effective than direct sowing of seed in woodland. Many small seeds of native trees, like birch, willows and poplars, are difficult to collect and difficult to store without rapid loss of viability. Large seeds, such as acorns and beech mast, are relatively easy to collect and to store but are so attractive as food for mice and squirrels that spot-sowings are liable to suffer severe losses even before germination. Sowing may also be complicated by the habit of many tree seeds to delay germination for a year (and sometimes more) depending on how they are kept after collection; some may germinate immediately if they are sown 'green' (like gean, if the stones are sown with the flesh of the fruit rotting on them) but go dormant for a year if they are allowed to dry. Some owners may be attracted to the idea of sowing seed directly in the woodland because of its 'naturalness', but it is seldom successful in practice. (See also chapter 14: 'Walnut'.)

CUTTINGS

Willows and poplars may be established from cuttings placed directly in moist woodland soils. Cuttings should be taken from vigorous one-year-old shoots, usually coppice; older material from the crowns of mature trees should be avoided. Cuttings should be about 25–30cm long with the top cut cleanly about 1cm above a stout leafbud and the bottom cut 1cm below a bud. (Cuttings of other species, including several conifers, are used for propagating in the nursery but not for direct establishment.)

BUYING PLANTS

Raising tree plants from seed can add interest to woodland work which may itself be valuable, but in most circumstances it is surer to buy plants from a reputable nursery specialising in tree-growing and much less costly to buy them there than from a retail garden centre; the range of species will be wider, the plant sizes more appropriate and the source of seed can be assured for most species. Most woodland planting is done with bare-rooted trees, grown in a nursery bed and then lifted before sale; container-grown plants offer some special advantages, although some kinds are more expensive.

Contrary to popular opinion, small tree plants are generally better than tall ones. You should look for a plant which is stocky, thick at the root collar, has a good balance between root and shoot and an undamaged root system (not chopped off or torn when lifted from the nursery bed) and one which has

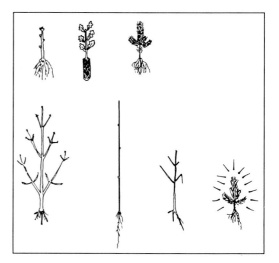

GOOD PLANTS AND POOR PLANTS
Top: good plants are well-balanced between root and shoot; stocky with a thick collar; full buds; compact fibrous roots; conifers have rich green foliage, not wilted.
Below: poor plants are ill-balanced, with large shoots and a small (perhaps chopped) root system; tall and lanky; broken twigs and roots; buds missing or weak; roots dry; conifers may be wilted, yellow foliage, heated in transport.

abundant fine root fibres. The shoots must be unbroken and not mildewed (a sign the plants have 'heated' in transport).

Tall, lanky plants, those which are mostly shoot on a small root system, ones which are thin at the root collar, or ones with roots dried in transport or mangled and pruned by bad lifting from the nursery bed should all be rejected. The reason for preferring small plants to large is not simply financial. It is extremely difficult to lift a large root system from the nursery without damage; consequently, large plants, unless specially treated, come with roots cut off (either deliberately or accidentally) and take a long time, perhaps years, to recover from the shock, while the leaves of the large crown make heavy demands for water and nutrients which the tiny root system cannot meet. A well-balanced small plant, undamaged, has a far better chance of survival and should soon outgrow the large plant. Small trees may be notch-planted, reducing costs. Table 4.1 gives some contrasting plant costs by size.

Bare-rooted trees usually spend one or two years in seedbeds and are then transplanted in the nursery for a further one or two years. (Plants should never spend three years either in the seedbed or in the transplant lines; unmoved for three years the root systems are bound to be so extensive that lifting damage is severe.) They are described as 1 + 1, 1 + 2, 2 + 1 and so on, the first figure being the number of years spent in the seedbed and the second the years in the transplant bed. Typically these trees are between 200mm and 400mm in height. The British Standard BS 3936 defines the minimum diameter of the root collar for plants of given height and species; this defines 'stockiness'. For example, a sessile oak 300mm in height should have a minimum collar diameter of 6.5mm, one 400mm in height an 8.0mm collar, and so on. You may ask the nurseryman to demonstrate this.

Seedlings are lifted and transplanted in the nursery in order to stimulate the development of a compact and fibrous root system capable of being replanted with minimum damage. Something of the same effect can be produced by undercutting the seedlings in the seedbed. Knives are pulled through the soil, a few inches below the surface and between the rows of plants, pruning the roots and causing them to branch. This is a quicker and cheaper method than transplanting and plants are referred to as 1 u 1 or 1 u 1 u 1. If the undercutting is properly done they should be as satisfactory as transplants.

In large-scale afforestation little use is made of container-grown plants apart from a few sensitive species. Corsican pine is frequently planted out as very young seedlings (only a few months old) in Japanese paper pots. Douglas fir, yew, holly and birch, which are tricky to plant as bare-rooted stock, are usefully grown in cells or plastic containers as plants up to 50cm tall. (Removed from their plastic containers, they are known as plugs.) For parkland, garden and small woodland planting a wide range of species may be used as container-grown or cell-grown stock. It is a serious fault when crowded roots spiral in a cell, a form of being 'pot-bound', since the contorted growth persists for years; 'rootrainer' cells generally avoid this.

Container-grown plants offer the great advantages of avoiding the constraint of planting only in the season when the plants are dormant, virtually eliminating the shock of transplanting since there should be no damage whatsoever to the roots and the cost of replacing failures should be greatly reduced or eliminated. The disadvantages lie in the greater cost, although this is now slight with modern plastic cells, and the weight of the material which has to be transported to the planting site. In very dry seasons, however, they appear to suffer rather more than well-

CELL-GROWN
SEEDLING
The rooting cell is 9 or 10cm long.

treated bare-root plants, perhaps because the roots fail to leave the peaty medium of the plug. When planted, the root collar should be just below ground level, with a little soil on top of the plug. For a few species, container-planting is almost obligatory, for instance eucalyptus. Most woodland planting is still done with bare-rooted transplants, although the convenience of the cell-grown makes these particularly attractive for small woodland work.

Note the difference between 'container-grown', which means that the plant has been grown in the container, and 'containerised', which means the tree was put into the container (with what effect on the roots?) at least five minutes before reaching the garden-centre shelf.

Orders for plants should be sent to a forest nurseryman as early as possible (by July for autumn or the following spring), specifying species, sizes, numbers required, the date required and the place for delivery. Stock in transplant beds cannot be lifted until plant growth has stopped and shoots have hardened. Modern nurseries now use cold stores to hold seedlings temporarily between lifting and transplanting and to hold transplants in a dormant state ready for despatch to customers; the technique usefully lengthens the feasible planting period.

When ordering plants from a nursery you should ask the provenance (the seed source) of each lot (see chapter seven); a reputable nurseryman will have no difficulty in telling you. The seed source is important. Tree species which grow over a wide area have, over thousands of years, developed sub-populations best suited to local conditions, especially of altitude and day-length. Physically they may appear identical, but they are physiologically distinct, for instance in the time of leaf-flushing, susceptibility to frost, length of growing period and so on. The purchaser should try to ensure that plants are grown from seed from the most suitable source. The *provenance* is the location of the tree stand where the seed was collected; in the case of a native tree that may also be the *origin*. In the case of non-native species the *origin* is where the original seed import came from. For instance, beech plants might be raised from seed with a provenance and origin in the Chiltern Hills, a native stand; European larch, on the other hand, might be of Morayshire provenance, the seed having been collected there, but of Sudeten, Czech Republic, *origin*, where the seed for the Morayshire trees came from.

Native species should be from seed stands as local as possible. You should not, by purchasing, support any nurseryman's unnecessary seed imports of species readily available in British woodlands. Plants grown from seed collected further south may come into leaf unseasonably and suffer from early and late frosts.

Within the European Union there are regulations, applying to 13 major tree species, which prescribe that plants for woodland planting should be grown only from seed collected from registered seed stands, the intention being to ensure that the stock is derived from good parents. The Forestry Commission maintains the register of stands in the UK and is responsible for seed testing.

If the nursery is reasonably close, you should visit it and ask to see the stock you require. A good nurseryman will provide sample plants which should be kept for comparison with those delivered later.

SEASON FOR PLANTING AND PLANT HANDLING

Bare-rooted trees may be planted in open weather any time after the growth has finished in autumn until growth starts in spring, usually from early November until March or April in the north, preferably before the start of root growth which precedes bud burst. In most districts there will be a mid-winter period when planting is impossible because of frozen soil and snow. A serious restriction in spring is the cold, drying, windy weather so common in March; on such days plants whose roots are exposed are quickly killed. Container-grown stock may be planted over a wider period, but the breaking of fragile growing shoots is a problem in the late spring.

As a general rule, deciduous trees should be planted before evergreens; broadleaves may be planted in the late autumn, followed by larches in late winter, pines, Douglas fir and spruces last. Wet and exposed sites should be planted in spring rather than autumn, in order to reduce the damage by gales loosening the plants and by desiccation. In low-rainfall areas many foresters favour autumn planting so that plants may establish before the dry spring. In small woodlands, however, with little pressure to complete the planting programme, spring planting is probably best.

When plants are received from a nursery they should be handled with care. It is good practice for trees to be packed in robust plastic bags, black inside and white outside, which offer good insulation from heat. Bags or bundles of plants should not be thrown off a truck or dropped from a height; such

treatment breaks the water columns inside the stems and the plants do not thrive thereafter. If the weather is frosty, the bags should not be opened but should be stored for up to ten days in an unheated but (if possible) frost-free building. If the plants cannot be used immediately afterwards, the bags should be opened and the plants kept moist. Take care they do not heat by fermentation.

Plants which have to be held for longer than ten days or so should be heeled-in (*sheughed-in* in Scotland) by digging a trench deep enough to take the roots, setting the bundles along the trench and covering the roots generously with firmed, moist soil.

Following the care taken to grow, transport and store them, it is careless to allow severe damage to be done immediately before planting. Bundles are opened and divided among the workers, perhaps on a windy spring day; the fine roots will dry up in only minutes if they are exposed and the trees will be severely shocked or even killed. The roots must be protected from sun and wind with a moist sack thrown over them until they are placed, root down, in the planter's bag which should contain a supply for not more than an hour's work in order to avoid drying. Ideal planting weather is windless, cloudy and drizzly. Avoid frosty days.

PLANTING METHODS

The method used for planting trees depends on the species, the soil, the size of the seedlings and, greatly, on the scale of the job. There are estate records of Scots pine seedlings being planted in sandy soils by women who averaged more than 2,000 plants each day, using planting 'daggers' – heavy iron spearheads on two-foot-long heavy wooden shafts – which were thrown into the ground a yard ahead of the worker. The trees were planted in vertical notches almost as quickly as the worker could walk, but such high productivity is feasible only with an 'easy' tree and on very light soil.

Most small woodland planting should be done with a spade or a mattock. For convenience the spade should be heavy, with the rather narrow blade in line with the handle; spades which are light and have a 'lift' on the blade are difficult to drive into turf or stiffly compacted soil. The mattock is now a much underrated tool in Britain although formerly widely used, as it still is in Continental Europe. In many respects it is more versatile and more efficient than a spade.

When planting bare-root trees in areas with a grass turf which has not been spot-treated with herbicide, the first action should be to screef a patch at least 50cm in diameter, in the centre of which the tree will be planted. The objective must be to place the tree firmly in the soil, so that the root collar is at ground level and the stem vertical, with the roots well spread out in their natural position in as good soil as the site provides. This can best be achieved by pit-planting. For this method:

- Dig a hole as wide as the root system and slightly deeper, keeping brown top-soil separate from poorer sub-soil.
- Loosen the soil at the bottom of the pit, put any screefed turf in the bottom, hold the tree in position and back-fill the soil dug from the pit, ensuring that the fine top-soil is shaken into the roots.
- Firm the soil when it is half-filled, ensure the tree is upright and at the correct height, then complete the back-filling and firm

PIT-PLANTING WITH MATTOCK
Dig a pit large enough to hold the roots in a natural position. Back-fill half the best soil round the roots and firm; check the depth and adjust if necessary. Replace the remaining soil and firm with the ball of the foot (not the heel, which scrapes the plant and compacts the soil).

well with the toe of your boot – taking care not to debark the tree – to ensure the roots are not dangling in air pockets.

Pit-planting should be used for all garden and park trees and is worth using in woodlands when the number of trees is modest because the extra care should be well repaid in the establishment and growth of the plants.

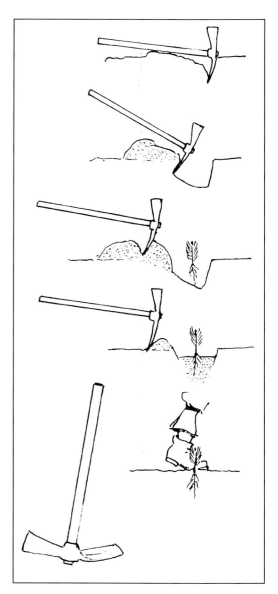

Poplar: a 12-year-old plantation of a single clone

Mixed broadleaves in an unmanaged farm woodland

Woodland for people: a walk in the famous Amsterdam-bos, the Netherlands

Farm woodlands: high-value sycamore in East Lothian

Design for stream management: tree-planting kept well back, allowing in sunlight and conserving natural tall vegetation (© Forest Life Picture Library)

Tree-planting for colour: a shelter strip at Battleby, Perthshire

Landscape with farm woodlands: the outline of the hilltop conifer blocks is too abrupt

Hybrid larch thinned to maintain deep crowns, Dunkeld, Perthshire

Sessile oak: a large-pole stand of the highest quality

Natural lowland forest of sessile oak, birch and juniper, Cawdor, Nairn

Sessile oak: group regeneration at small-pole stage, Lethen, Nairn

High quality and high value: mixed woodland of ash, larch and Douglas fir in a sheltered valley, southern Scotland

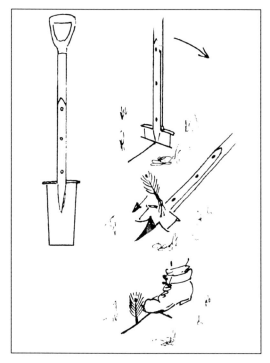

PLANTING SPADE AND T-NOTCH
Cutting into tough soil is easier using a narrow spade with no 'lift', but a garden spade will serve.

VERTICAL NOTCH PLANTING
Below are the steps in vertical notch planting. Above is a Schlich-pattern planting spade, suitable for this notching, and two planting spears, a nineteenth-century spear locally made and used around the Moray Firth and a modern spear designed for cell-grown stock.

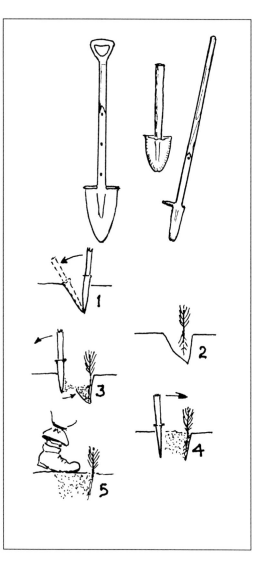

In large-scale afforestation the planting of bare-root trees is done by notching. In either L or T-notching the spade is first driven into the soil to make the vertical cut; to form the base of the L or the cross of the T, the second cut is then made at right angles and the spade is levered back to open a flap or flaps. The plant is then inserted into the vertical leg of the notch and the roots are spread out beneath the flaps. The spade handle is then raised to close the flaps, which are pressed firm with the boot.

In another notching method the spade is driven in vertically and levered to open a slit, into which the plant roots are dropped. The vertical notch is then closed by driving the spade in parallel to the first line and levering the first notch shut, trapping the roots; the ball of the foot is then used to firm the planting and close the notches. (Using the heel, there is a tendency to scrape the tree.) Cell-grown plants can be notch-planted with

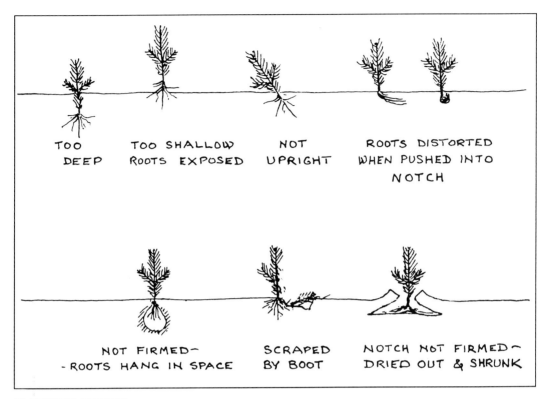

PLANTING ERRORS

a 'spear' in a similar fashion.

Notching with the mattock is done in the same way but is perhaps rather easier and faster than with the spade. The cross notches of the L and T are made at the farther end of the vertical, relative to the planter, whereas they are at the nearer end with the spade.

In large planting programmes, the attraction of the notching methods is their speed. An experienced person may plant 800 or 900 plants a day notching, but only around 300 pit-planting. The principal shortcoming of notching is the poor distribution of the roots, which at best are sandwiched vertically with vertical notching or horizontally with L or T-notches. All too easily, however, the roots are bent into a U-bend or crammed into a tight ball, requiring the tree to grow a new root system before it can begin to put on

shoot growth. Notching is satisfactory if the soil is a sandy loam but not so if it is a silt or clay where compaction of the roots is more severe. Pit-plant if you can afford the time.

For planting on wet sites, turfs may be cut as slices from a drain and spread at the chosen planting distance over the site, or they may be cut at each planting site and inverted immediately alongside. (see diagrams pp. 121–2). The larger the slice of turf, and thus the more rootable soil for the tree planted in it, the better the tree should grow. It should be at least 300mm square and preferably at least 150mm thick. The tree may be notched into a thick turf; in the case of thin turfs the notch will cut right through and the tree roots will be spread in the 'sandwich' of vegetation. For success, the turf must be firmly settled on the soil surface, otherwise roots will dry or the turf

STAKING TREES

The swaying of a tree in the wind stimulates the growth of thickness in the stem. Staking results in a thin stem so that, when the stake is removed or rots, the tree may fall over. Staking attracts vandalism and makes the tree easy to snap. If you must stake, use a short post and remove it soon.

itself, as it dries, may roll over. Wiry vegetation such as heather will keep a thin upturned turf airborne and should be burned off first.

In normal woodland practice trees are not staked, and well-proportioned trees of 300–400mm in height planted in gardens do not require staking if they are properly planted. Large standard trees with pruned roots planted in parks and streets will probably require staking because of their poor root-shoot balance, but it should be minimal and as short-term as possible. Use a stake less than half the height of the clear stem below the branches or less than one-third the height of the tree it is to support. Stake support relieves the stem of stress and thus deprives it of the stimulus to thicken, so the staked tree may grow too heavy to hold itself up naturally when the stake is removed or rots away. Dig the pit large enough to receive the tree roots,

drive in the stake in the empty pit and then plant the tree in the normal way and tie it to the stake. But in almost every situation it is better to use a small plant and no stake – the wild forest had no stakes and the trees were better than in our plantations. Large staked trees are a challenge to vandals and the tie provides the convenient place to break the stem. If a tree has to be staked, the tie must not strangle the stem, and the tie and stake must be removed as soon as the root is sufficiently secure to hold up the tree.

Occasionally fast-growing young trees in soft ground are blown about so severely in winter that the bases form open sockets in the soil at the root collars. (Douglas fir is so prone.) The sockets fill with water, softening the soil even more and exacerbating the problem. Firm fresh soil in the socket with the boot; sometimes the tree will require a temporary prop.

FERTILISERS

Except on the most infertile soils, such as industrial wasteland, moorland and heather areas being planted with spruce, manuring should not normally be necessary.

On infertile sites such as ironpan and peaty podzols, deep peats, etc., especially those with heather, spruce may require 35gm (a small coffee-cupful) of ground mineral phosphate to be sprinkled round each tree at planting time and again in about year eight. In heather areas spruce tends to 'go into check' (indicated by foliage turning yellow), when it remains alive but does not grow appreciably for several years due to mycorrhizal antagonism and nutritional deficiency. On the most infertile soils, potassium may also be required, as may nitrogen, especially on deep peats. It is unlikely that the small woodland manager would be planting these sites, so such fertilising might arise only for some farm shelter-planting; in these cases, local professional advice should be taken on the fertiliser required. Financially this is well justified to save an unnecessarily large application or to save the failure of the planting. Only pure spruce crops are likely to require nitrogen fertilisers. On difficult sites pine and larch are better able to obtain sufficient nitrates than spruces, and research has shown that when Sitka spruce is planted in mixture with larch or pine on heathery sites, it enjoys a much better nitrogen supply at second hand, as it were, from the other species – a kind of 'alternative therapy'.

On industrial wasteland, full NPK fertiliser applications may be required and both a soil analysis and professional advice should be obtained before spending money. All granular fertilisers should be applied as dressings to the soil surface, not dug into the planting site, where they may damage the roots.

An alternative to mineral fertiliser may be the application of dry cake sewage sludge, which offers an environmentally friendly way of using what was, until recently, a serious pollutant of coastal waters. On industrial wasteland, 150–200 tonnes per ha may be applied before planting to supply much-needed organic matter. The material is innocuous and its use is an important aspect of modern forestry (mostly the larger units, but not necessarily so). Sludge may be applied to young stands on most soil types.

Again, the planting of alders, whose actinomycete micro-organisms can fix atmospheric nitrogen, provides an alternative to the fertiliser bag, helping the alders themselves and neighbouring trees. Do not apply nitrogen fertiliser to alders. Tree growth can be greatly improved for other species by importing some forest litter to the planting sites, thereby inoculating the soil with forest micro-organisms which, as mycorrizas, can obtain nitrogen from decomposing leaf-litter.

BEATING-UP

Some young trees are almost bound to die in the year after planting as a result of rabbits, deer, voles, summer drought and so on, even if planting has been done expertly. With plants spaced at 2m or less, it is not necessary to replace every tree that fails and up to 10 or 15 per cent failure evenly spread may be accepted. If there is a patch of failures they will require replanting, or *beating-up*. This can best be done in the autumn after planting; replacements should be pit-planted to give them every chance to catch up growth.

Failures in the second and subsequent years after planting pose increased difficulty because the replacements will develop behind the original planting and late beating-up extends the period when weeding is necessary. After two years, replacement is not worth attempting, apart from large patches killed by fire or a similar disaster.

Beating-up is expensive. It is far better to

plant with such care that the original planting succeeds well, and to tend and protect it so that failures are negligible.

WEEDING

Weeds close round the small trees compete for nutrients and moisture, shade them from sunlight which they need for growth and, in the autumn, tend to collapse upon them like a rotting wet blanket. Weeds which will pose such a serious threat to the trees should be dealt with before planting, by crushing, herbicidal treatment, mounding and screefing.

Formerly forest practice was to cut weeds growing round the tree plants, generally using a reaping hook or a small scythe. Experiments have shown, however, that cutting weeds does little or nothing to improve the trees' supplies of nutrients or water, while inevitably a proportion of trees are cut accidentally. Do not try to control weeds by cutting them.

On upland sites where trees are planted on mounds or turfs the reinvasion of weeds may be sufficiently slow that weeding will not be required. On lowland sites, even with mound-planting or screefing, weeding will be necessary, especially on fertile heavy soils in south-west England and Ireland, where two or three years of weeding is usual. On grassy sites a pre-planting treatment should create a grass-free planting area at least 80cm in diameter. After planting, if weeds have invaded strongly, they should be treated with a herbicide (a granular form or a 'weed-wipe' are probably most convenient and safe) or trampled back. Breaking the weeds with booted feet is preferable to cutting.

Careful planting and giving the trees a weed-free 'holiday' in which to establish themselves and begin to grow is the secret to minimising weeding costs. Poorly planted trees, especially broadleaves, which suffer heavy weed competition from the start grow so slowly that, in spite of continual attention, they are still struggling to get above the ground vegetation years later.

Do not over-weed merely to make the planting area tidy. Once the trees have their tops clear of the surrounding vegetation they should be safe. Vegetation in the spaces between rows is desirable in that it reduces exposure and the risk of desiccation and encourages earthworms, which are the basis of the nutrient recycling system.

MULCHING

This practice provides control of weeds and a means of enhancing the growth of young trees, especially specimen trees in parks and

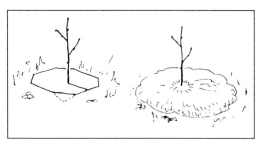

MULCHING – MATS OR ORGANIC
Mulching mats must be secured or they would blow away. Stones, soil or turf may be used; polythene sheet may be held in place by thrusting the corners into the soil with a blunt spade.

gardens. Mulching is impossible in extensive forestry but has high value in establishing wide-spaced poplar on rich sites. It may be done with almost any organic matter or with mats of polythene and similar materials. Organic matter commonly used for mulching includes leaf-litter and leaf-mould, rotted grass mowings, waste farm straw, turf and tree bark chips. Light material which can be blown about by the wind is generally unsuitable, as is sawdust, which leaves the soil short of nitrogen. The mulch should be spread around the tree in a circle up to 1m in diameter and

100mm deep, preferably leaving the tree stem clear, as a protection against rotting.

Proprietary mulching mats may be purchased, and less satisfactory ones can be made from old fertiliser bags on a farm or from old bitumen roofing felt. The mat should be about 0.8–1.0m², with a slit from the centre to one side, to allow the mat to be slipped around the planted tree. It must be held in place, most effectively by pushing the corners with a spade into slits in the soil, otherwise with stones placed on the edges. Weed suppression and soil moisture conservation are achieved by both the organic and sheet material mulches. The cheapest proprietary mats are about £0.40 each, the dearest £2.40; proprietary organic mulches cost about £2.00 a tree (1997 prices).

TREESHELTERS

Much use is made of plastic tube treeshelters and netting guards in small-scale tree-planting where there is a risk of browsing damage. Their use will be discussed in chapter eight on protection of woodlands and trees.

STUMPING BACK

In a few broadleaved species, of which oak is the most important, it is sometimes useful to cut down the plants to 7 or 8cm above soil level when they are established at two or three years old. The object is to replace a poorly developed and bent shoot with a straight single one from the cut stool, which oak does very vigorously. A proviso must be that the weed growth can be handled, so that the plants will not be damaged a second time.

ESTABLISHMENT

A plantation may be considered to be established when no further beating-up or weeding is required. There will follow a period

when woodland work will be restricted to maintenance of the drains, so that they do not become blocked with debris, and to general protection, principally against browsing animals and fire. New planting, restocking and acceptable natural regeneration are eligible for planting grants from the Forestry Authority, from whom details and application forms may be obtained. An outline of the types of grant currently available is given in appendix 1.

FURTHER READING

Edwards, C., Tracy, D.R. & Morgan, J.L. (1993), *Rhododendron Control by Imazapyr*, Forestry Commission Technical Paper 3, HMSO, London

Forestry Commission (Ed. B.G. Hibberd) (1991), *Forestry Practice*, Forestry Commission Handbook 6 (11th ed.), HMSO, London

Forestry Commission (1995), *Mulching Trial: Evaluation of Currently Available Mulching Systems*, Forestry Commission Technical Development Branch, Ae Village, Dumfries

Tabbush, P.M. & Williamson, D.R. (1987), *Rhododendron ponticum as a Forest Weed*. Forestry Commission Bulletin 73, HMSO, London

Taylor, C.M.A. (1991), *Forest Fertilisation in Britain*, Forestry Commission Bulletin 95, HMSO, London

Willoughby, I. & Clay, D.V. (1996), *Herbicides for Farm Woodlands and Short-Rotation Crops*, Forestry Commission Field Book 14, HMSO, London

Willoughby, I. & Dewar, J. (1995), *The Use of Herbicides in the Forest*, Forestry Commission Field Book 8, HMSO, London

Silviculture:
The Art of Growing Woods Well

6

The forester, while concerning himself with obtaining from the forest the most and the best possible for his fellows, must also see that this is done without destroying, or even damaging, it.

Mark L. Anderson, 1956
BBC Third Programme Radio Broadcast: *Time for Forestry*

Silviculture is the husbandry of woodlands, a branch of applied ecology and the very foundation of sustainable forestry practice. Sound silviculture is the only way in which woodlands can be financially sound; it is the basis, not the antithesis, of woodland economics.

There are three main types of woodland based on how the trees are regenerated (or are intended to regenerate): high forest, coppice and coppice-with-standards.

HIGH FOREST

High forest is regenerated from seedlings, whether by planting, natural regeneration or direct sowing. The term does not refer to the height of the trees, which may be young or mature. This is the most usual system of silviculture in Britain today. It can be run in one of the following ways:

- Clear-felling and replanting, which is commonest in Britain because it allows the lowest-cost harvesting and because of a wish in recent decades to change the tree species of the forest.
- As a shelterwood system with natural regeneration, perhaps reinforced by planting, so that after a period of progressive thinning of the mature stand to increase the light on the ground, a crop of seedlings creates a new, more or less even-aged stand.
- As two-storeyed high forest, when a middle-aged stand, generally of a species casting light shade such as oak or larch, is underplanted to create a shade-tolerant understorey, say of beech.
- As clear-felling areas somewhat reduced in size and arranged to form a strip or group clear-felling system.
- Groups, further reduced in area and irregularly arranged so that a full range of age classes is present within the same wood, in a group-selection system.
- Stems of all sizes (and often more than one species) may be closely mixed together in a stem-by-stem selection system.

The arrangement adopted depends on the long-term objectives set for the woodland and equally on the characteristics of the trees there. The manager of an extensive forest who wishes to minimise the costs of harvesting a bulk crop for a pulpwood market is sure to be

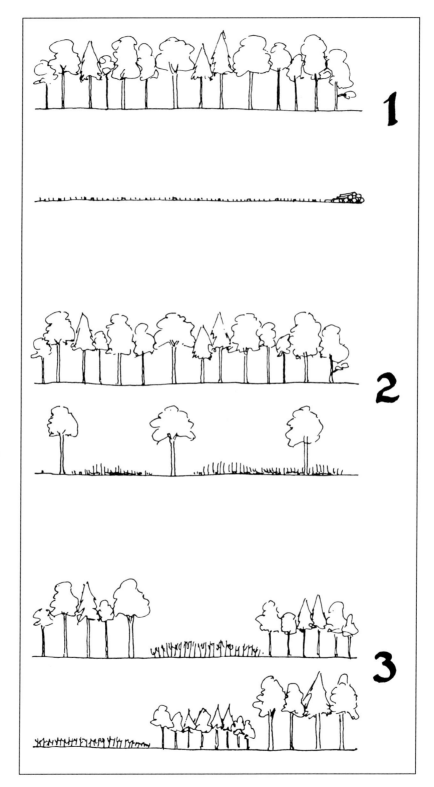

attracted to machine-harvesting and hence to a clear-felling and replanting system. In contrast, the manager of woodland heavily used for public recreation and valued for its wildlife conservation is likely to favour a system which provides nearly or completely continuous tree cover, such as group clear-felling, group-selection or even stem-by-stem selection.

The silvicultural characteristics of the trees forming the woodland set constraints on its husbandry. Stem-by-stem selection is possible only with species which are fully shade-tolerant, such as beech and silver firs. Furthermore, a stem-by-stem selection system is increasingly difficult to work as one moves north, because of the lower angle of the sun and the poorer illumination of the forest floor in small gaps. With less shade-tolerant species, continuous woodland cover can be achieved only with a group structure, and in the north the group size must be quite large in order to provide the light needed for the saplings of light-demanders like oak or Scots pine to grow well (see chapter seven). The cutting of a mature tree releases the smaller trees growing alongside and allows germinating seedlings to establish themselves.

Two-storeyed high forest may be intended for ecological efficiency, to use the light 'going to waste' under the thin canopy of oak or larch; or it may be intended to shade the stems of the overstorey trees, to minimise the growth of epicormic branches which develop commonly on the boles of pedunculate oak in middle age; or it may be an attempt to raise the intended replacement crop under the maturing overstorey, to 'telescope' the rotations. This last concept is admirable but the execution is difficult. For instance, if larch, thinned to about 250 stems per hectare, is underplanted with Douglas fir, the larch must be cut before the leaders of the understorey grow into the larch branches and become damaged there, and harvesting the larch inevitably does some damage to the understorey: very little if they are still wiry seedlings but a great deal if they are small poles.

These systems all have in common the fact that they are founded on the stands being regenerated by seedlings.

COPPICE

In contrast to high forest, coppice is based on the trees regenerating vegetatively from stool shoots or suckers arising after the parents have been cut. Each division (or coupe) of the

HIGH FOREST SYSTEMS

1. Clear-felling and replanting: simple in harvesting and management, and no delay, but the shelter of the woodland is lost.

2. Shelterwood system with natural regeneration: a seeding felling removes all but the seed trees which are progressively removed as the regeneration becomes established.

3. Group clear-felling: several rather large groups of various ages are present in the woodland but not necessarily all the ages in the rotation. The woodland environment is fairly well conserved.

4. Group-selection system: the woodland comprises groups, each of one age, covering the full range of ages in the rotation or production period; the size of the groups depends on the shade-tolerance of the species, but they tend to be rather small in order to provide good shelter and conserve the woodland environment.

5. Stem-by-stem selection system: requires high skill to manage and must be restricted to fully shade-tolerant species, e.g. some silver firs and beech.

6. Two-storeyed high forest: usually a temporary condition as a step to replacing the main stand or to control soil erosion etc. Difficulties increase with the age of the understorey, especially if it grows into the crowns of the larger trees.

COPPICE WORKING

wood is cut over at a set number of years, depending on the species and the type of produce required. A large wood may be divided into the same number of coupes as there are years in the rotation (the production period), so that one coupe may be cut and the produce marketed each year. Thus if a 6ha wood of mixed coppice is being grown to an age of 24 years, the wood should be divided into 24 coupes (or cants), each of 0.25ha. In this way the same amount of produce is marketed each year, the workload is the same and butterfly habitat is sustained, as is pheasant cover, employment and all the rest. This is a fundamental of traditional forest management: sustention of the yield promotes sustention of all the other outputs and the inputs; it delivers sustainability.

Coppice management was applied over hundreds of years in most of Britain and Continental Europe, supplying wood for country industries, for firewood and latterly for emerging manufacturing industry. Coppicing began with the repeated cutting of natural woodland but, with increased demand for its products, many new woodlands were planted for coppice working. Few coppices are now managed systematically, except sweet chestnut for cleft paling.

Hazel coppice was used primarily for the manufacture of farm hurdles for fencing, but also provided spars for thatching and other products. It was cut generally at eight years.

Ash coppice provided a wide range of handles (scythes, hammers, axes), turnery (bobbins, chair legs, etc.) and wood for wheel-making, with rotations between 12 and 30 years. Beech and hornbeam coppice was used principally for chair-making, turnery and other furniture, oak for tan-bark and to make charcoal for metal-smelting, sweet chestnut for fence-making, and so on. In many instances the end product was manufactured in the forest, by chair-bodgers, hurdle-makers or charcoal-burners.

The coppice stools should be spaced about 3m apart, 3.5m for oak to 2.2m for hazel and other short-rotation species. Thinning the shoots on each stool was sometimes done to improve the stem quality but is not usually necessary or desirable. Cutting can be done with a saw but an axe or heavy bill was considered better for light material, in order to leave the stools clean-cut and less liable to absorb water and rot. Young shoots are very palatable to roe deer and rabbits (and to farm stock if they can gain access); protection for two years may be necessary, but a large area of vigorous coppice tends to grow faster than a few animals can eat it, so fencing against wild animals may be unnecessary. Farm stock must always be excluded.

The markets for coppice produce weakened progressively from about 1860 to 1930, and with their collapse, not least that for firewood, almost the whole practice of coppice management has been lost in the United Kingdom, Ireland and most of Continental Europe. This represents a sad loss both of country crafts and of specialised wildlife habitats. It is ironical that many of the substitute products are technically inferior and non-renewable. The only traditional coppice in good commercial management in England is sweet chestnut in southern counties, marketed for fencing and poles, although interest is now reviving in working other species.

When coppice working is abandoned, the stools of several species appear markedly to lose vigour; hazel, for instance, weakens to produce straggling shoots which are quite different from the robust wands of regularly worked coppice. The problem arises of how abandoned coppice may be managed. Many areas have been cut over and replanted as high forest, often of conifers. Other areas may be converted to a longer rotation by singling the shoots to one on each coppice stool and then treating the stand as if it were high forest; the result is called *stored coppice*. Timber trees from stored coppice are regarded as poorer than those from *maidens* (i.e. seedlings), mainly because of the shape of the base and the higher risk of rot entering the stem from the stool, but they are vigorous and have some years' growth advantage compared to cutting and planting a new crop. Storing has some obvious limitations; the new crop is inevitably the same species as the coppice; ash, oak, sycamore and sweet chestnut generally give satisfactory results but hazel will not and beech is unreliable. Furthermore, there must be sufficient stems of good quality to make storing feasible, a problem if the stools are more than 3m apart. It is usually impossible to fill gaps with new planting since the existing shoots quickly overtop seedlings and suppress them. With very patchy crops it is probably better to cut the stools back, treat with herbicide and replant, on a group clear-felling system based on the coppicing coupes.

There is now renewed interest in short-rotation coppice which may be grown as a crop to supply sustainable bio-fuel or raw material for the manufacture of panel products. Willows and poplars may be grown on a rotation of three years, closely planted in lines for mechanised harvesting. This system of cropping is feasible only on fertile arable land and in a mild and kindly climate. The land must be reasonably level, to allow machine-harvesting. This is not a system to be adopted casually; you must know there is a market for the produce and take expert advice on the most suitable combination of cultivars for the locality (see chapter 14).

COPPICE-WITH-STANDARDS

In some parts of Britain, notably in the Midlands, a common woodland-management system was coppice-with-standards. Trees were grown for timber in a matrix of coppice beneath them, thus providing in a single woodland both heavy constructional timber and the range of coppice goods and firewood for local use. It was the multi-purpose system *par excellence*. The standards and coppice might be the same species, for instance both oak, but more frequently they were different: oak standards over hazel or ash or sweet chestnut or lime, or a mixture of several of these. The system called for skill to ensure that sufficient recruits were present to provide a choice of future standards and also to ensure that the standards were not so numerous as to suppress the coppice. The usual rule was that the crowns of the standards should not occupy more than one-third of the area and, in any event, their crowns should not touch.

In a wood of ash coppice with oak standards, it might be that the ash would be cut on a rotation of 20 years. At each cutting of the coppice the mature oak of 120 years would be cut, unwanted younger standards would be thinned and replacement standards would be planted, as well as ash required to replace worn-out coppice stools. Deaths of stools may amount to 1–5 per cent at each cutting, and these must be replaced.

In fact, there could be flexibility in the management of the mature standards. A particular tree could be left to grow for an additional coppice rotation, or more, and a stem which was not promising well would be cut early. The proportion of the various ages

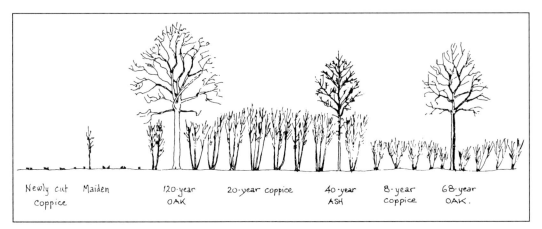

COPPICE-WITH-STANDARDS

The age of the standards is generally a multiple of the surrounding coppice. The shade of the standards reduces the growth of the coppice beneath the crown, so too many standards reduces the coppice yield.

had to be sufficient to ensure future recruitment. Approximate numbers per hectare of the various classes of standards might be as follows (remembering that judgement and the species of standard and coppice affected the result); standards might be of more than one species:

Newly planted standards	56
20 years old	28
40 years old	16
60 years old	8
80 years old	6
100 years old	4
120 years old and now being felled	3

Standards of each age-class would be removed at each visit, cutting the poorest of the immature, retaining the best towards maturity. The form of the standards is easy to recognise in these old woods, boles clear of branches up to the top height of the coppice at its rotation, the crown then spreading wide above that point.

The woods of this structure are still a notable feature of the Midlands and even north to southern Scotland. The system is particularly sympathetic to wildlife, combining the special habitats of coppice and those of the old trees of high forest, and there may be a place for restoring this management system in some small woodlands. Owners who wish to promote multi-purpose woodland and who have a use for coppice wood should seriously consider this system of management which offers so much interest and flexibility.

STAND DEVELOPMENT

Common to all the woodland systems is the basic ecological characteristic of trees: that they are designed to dominate other vegetation. Silvicultural practice should promote this as quickly as possible, eliminating weed growth which competes for water and nutrients. If there is a need to have areas of open ground for shooting, as wildlife habitat or for visual amenity, it is better to set aside areas to be open glades in a matrix of proper woodland. (Widely spaced trees over the whole area as in parkland may offer high amenity values but generally provide low-value timber. The problems of agroforestry, which also has trees widely spaced, are referred to in chapter 12.)

It is useful to recognise six development stages in the life of the tree stand: seedling, thicket, sapling, small pole, large pole and timber. The seedling and thicket stages have been described in chapter five. After successful establishment there is a period, which may be only five or six years or may be as many as 25, in which work will be confined to maintenance of drains and fences and general protection, notably from fire but perhaps also from browsing animals. Thereafter, as the stand of trees develops into saplings and small poles, the woodland manager has to show skill in balancing competing pressures.

On the one hand there is merit in keeping the stand rather dense, especially broadleaved species, so that side branches remain small (and may be shed in most species) and the main stem grows tall; so that the canopy is complete, capturing a high proportion of the incoming radiation from the sun; and so that there is a good number of trees from which the mature crop can be selected.

On the other hand it is necessary to ensure that the trees, at least those destined to become the main crop, have sufficient space to develop a balanced crown and to remove damaging, or potentially damaging, trees at an early stage, before the main stand has been ruined. This is especially true of coarsely branched, misshapen trees, sometimes called 'wolves', which may be the result of their genetic make-up or of some early damage. Also to be removed are stems of fast-growing species not required in a timber crop, such as willow or excessive numbers of birch in a thicket of oak.

It may be objected that, from a purely ecological standpoint, these interferences are unnatural: if the willow overwhelms the oak, so be it. This argument would be strong if there were an abundance of lowland oak forest maturing and regenerating well, but unfortunately this is not so. There are few relics of the lowland broadleaved forest in robust condition in Britain. In an extensive forest hundreds of patches of willow and birch seedlings could be accommodated without concern; the oak could afford to wait 50 or 60 years before regenerating in the then failing willow patches. But, in a small area of oak woodland being invaded by willows, will there still be oak there 60 years hence to get the chance of bouncing back? And what will pay for maintaining the woodland while waiting for this recovery?

In general the intense competition among seedlings is gradually resolved, with the strongest and fittest emerging as dominants which suppress their neighbours and establish their right to parent the next generation. Man sees advantage in harvesting the suppressed and himself selecting trees to form the final stand which will have the characteristics of species, shape, timber quality and so on which best suit his needs.

CLEANING

This is the first operation after weeding has ceased and the new stand is established. In plantations it involves cutting unwanted invading trees; in natural regeneration it may also involve respacing plants of the main species, more especially of conifers. Nevertheless such respacing can easily be overdone; wide spacing allows heavier branching and greater development of juvenile wood (see chapter 16), a major defect of conifers for construction-timber markets. The operation produces no saleable material and perhaps not even small firewood. In dense natural regeneration and closely planted stands, timely cleaning is important, followed by light and frequent thinning to maintain the balance between growing a vigorous crown and the competition necessary for good tree form; this applies especially to broadleaves and to conifers such as larch.

BRASHING

This involves the removal of dead side branches to a height of about 2m to allow access to the stand. It is now very seldom attempted in low-input big-time forestry, but it is well worth doing in high-value woodlands because it allows freer access and higher-quality work, especially thinning and pruning. Do not use an axe, hedging knife or billhook for this operation – the result is always damaged stems – nor a chainsaw, which is even worse. Use a curved pruning saw on a pole, cutting on the pull-stroke (much more efficient than a narrow-bladed push-stroke saw which is sometimes offered for the job). There is no point in brashing suppressed and dead trees; ignore them and they will be knocked down later. Do not brash the outside row of trees in a young pole plantation, especially in a windy place; the branches help to maintain the sheltered woodland conditions in the stand.

PRUNING

Whereas brashing is intended to improve access to the stand, pruning aims to improve the timber product. It has two goals:

- To improve the form of the tree, removing double leaders and encouraging the development of a straight undivided stem, the latter especially in broadleaves, since this is the usual form of conifers.

- To produce lower stems (and hence the fattest log of each tree) with no branches and, therefore, a large amount of clear timber (this applies especially in conifers, since broadleaves more readily shed side branches).

In small woods, pruning should begin at the seedling stage with the removal of any double leaders which have developed after planting, using a pocket knife or secateurs. When broadleaved trees are grown densely the side branches which are deprived of full light by the upward-growing crown die back and soon fall off, provided they are so small that no heartwood has formed in them; for this reason the stand should be kept fairly 'tight' at least until the stems are clear of branches to the height of the required butt log. In contrast the dead side branches of the common conifers are more persistent, even those of such light-demanding species as larch and Scots pine, and consequently their capacity to produce clear timber naturally is less.

If pruning is to be cost-effective, it must be done early in the life of the tree. Most side branches grow from the buds on the leading shoot and as long as the branch is projecting from the stem there is a knot buried in the wood below. It follows that 'clear' timber (knot-free wood) can begin to be produced only after the branches have been cut off or have fallen naturally; the tree stem at the point of pruning should be no more than 12cm in diameter (10cm is better).

In order to get a log with valuable clear timber, the pruning must be taken up to 5–6m, and this will be possible only over several years, perhaps five. In conifers only dead branches should be cut; cutting live branches, especially of spruce, tends to induce undesirable growth around the wound, for instance producing 'pappy spruce'. In broadleaves the cutting of live branches appears to produce no bad reaction, although if overdone it would reduce growth.

Pruning broadleaves is more likely to be economic than pruning conifers, since the financial premium for high-quality veneer butts is high. It should be aimed at the trees likely to form the final stand towards maturity. There are two possible exceptions to the proposal that it may not pay to prune conifers: Scots pine and Douglas fir. Scots pine is apt to have loose knots and powder knots if dead branches are left, both of which

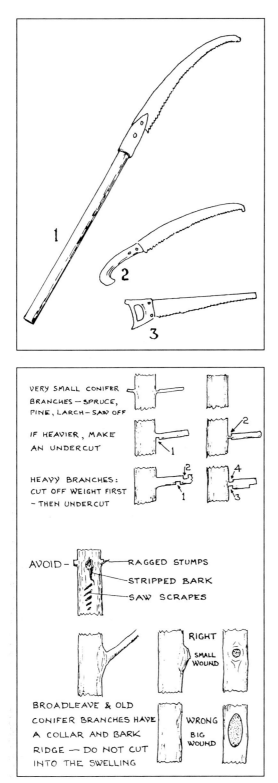

VERY SMALL CONIFER
BRANCHES — SPRUCE,
PINE, LARCH — SAW OFF

IF HEAVIER, MAKE
AN UNDERCUT

HEAVY BRANCHES:
CUT OFF WEIGHT FIRST
— THEN UNDERCUT

AVOID — RAGGED STUMPS
STRIPPED BARK
SAW SCRAPES

RIGHT
SMALL
WOUND

BROADLEAVE & OLD
CONIFER BRANCHES HAVE
A COLLAR AND BARK
RIDGE — DO NOT CUT
INTO THE SWELLING

WRONG
BIG
WOUND

PRUNING SAWS

1. The pole saw cuts on the pull stroke and uses two hands. The shape of the saw, ideally a parabolic curve, is critical so that the saw 'bites' into the wood progressively and stops before pulling through and out of the cut. This design feature is especially important when it is mounted on a long pole for high pruning.
2. A one-handed saw cutting on the pull stroke. It is much inferior to 1 above and very tiring to use.
3. A narrow-bladed saw working on the push stroke and designed for one hand only. Inefficient and exceptionally tiring, it should be rejected.

BRASHING AND PRUNING (below left)

cause the price of otherwise excellent stems to be severely depressed. In a woodland aimed at producing high-quality timber, prune sufficient pine to form most of the final crop trees, say 200 per hectare; the same can be done with really promising stands of Douglas fir, pruning perhaps 150 of the best stems per hectare. This is not standard advice, because the present price premium for pruned logs may not convince careful accountants that the operation is justified; it is speculative, on the hunch that in 50 years or more connoisseurs will pay high prices for outstanding material which, by that time, will be a very rare product indeed. The fact that big-time forestry does not prune at present is a signal to the manager of small woods that this is a fair speculative investment. Furthermore, it is work which can be done in a quarter-hour of spare time, turning leisure into an investment with a financial pay-off. One well-known English landowner used to (and may still) invite his house-guests to prune his trees before dinner.

THINNING

Apart from the choice of species to plant, no operation is likely to be so influential in obtaining a good result from woodland work

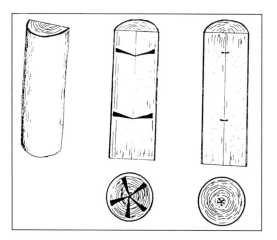

CLEAR TIMBER AND KNOTS
*The outside of the tree (left) may be clear of branches
and suggests that the timber may be clear of knots. But
if the stem has been pruned late in life, the wood will
be knotty and the work useless. In order to give value
in clear timber, pruning must be done when the stem is
small, then only the core of the mature tree will be
knotty.*

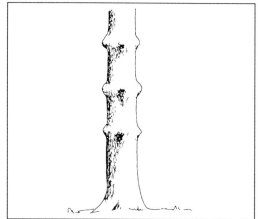

PAPPY STEM OF NORWAY SPRUCE
The result of pruning green branches.

as thinning. The aim in thinning is to remove a proportion of the stems so that the remainder may grow better. They achieve that by finding they have less competition below ground – more soil to exploit with their roots, and more space in the crown – so that their increased leaf-area intercepts more sunlight. As a result of thinning, the woody growth of each hectare is concentrated in fewer stems which are stouter than they would have been had the thinning not been done. The effect is to increase the proportion of sawlog-sized trees of the highest quality and to achieve maturity earlier.

In big-time forestry there is a tendency to do no thinning at all (harvesting mostly pulpwood at about 35 years) or to do systematic or mechanical thinning, first removing, say, every seventh row entirely, then an intervening row. This, however, has no place in small woodlands where thinning is best done selectively; the latter involves more

work but gives a better and more windfirm result.

The operation should aim to remove diseased, misshapen and unthrifty trees, especially those which threaten more promising ones. Although the principal aim is to enhance the condition and prospects of the stand which is to remain, thinning can also provide an intermediate income for the owner. In large, extensively managed forests, managers employ contractors both to mark and to cut the thinnings, which may provide a profit of only a few pounds per tonne for small roundwood. Consequently there is an incentive to avoid early thinnings when the stems are small and less valuable, and to increase the intervals between thinnings so that the volume to be cut is substantial ($35m^3$ to $50m^3$ per ha per visit in conifers). These demonstrate the watershed between big-time forestry and more intensive silviculture. Extensive forest management favours wide spacing at planting not only to cheapen the initial cost but also to avoid early thinning; it accepts a loss of quality in the remaining trees for the final crop in order to release a greater volume in the thinnings and thereby increase the revenue of each. These influences should be absent from small woodland work, which should aim at high quality, a different result with different practices.

It is useful to think of selective thinning in two stages: thinning for improvement and thinning for growth, although the two merge and neither purpose is entirely absent at any time. In the phase principally for improvement, the work involves mostly the removal of poor stems, especially those which are gross and crooked. In the growth phase, the forester should continually try to give the trees of the future mature stand the space which will allow them to grow well.

Regular thinning of most species (as distinct from cleaning and respacing as Christmas trees and greenery plants are removed) may begin when the fattest trees in the stand are about 8m tall. On fertile land, fast-growing species such as hybrid larch and Douglas fir may achieve that at about year 14 and the cycle of thinning may be three years. Slower-growing trees on poorer land may reach 8m only about year 30, and the appropriate thinning cycle may be six years or more. If trees have been planted at 3m spacing, thinning will probably begin later, say at 10m top height.

The frequency of thinning depends on the productivity of the site, the species, the age and rate of growth of the stand, exposure to wind and other factors, including the market for the produce. In general, young stands should be thinned on a cycle of three to five years (three if they are fast growers) and older stands every five to ten years. Broadleaves usually require thinning on a cycle of five to ten years, lengthening to 12 or 15 years near maturity.

In order to achieve continuity of purpose in successive thinnings you may choose to identify potential final-crop trees very early in the rotation. After the cleaning, you may mark a number of well-distributed trees from which the final-crop trees are likely to be chosen, for example with paint bands on the stems. The selected trees might usefully be about 7m apart in oak or Douglas fir, about 5.5m in ash, sycamore and other broadleaves, and about 4.5m in pine, spruce and larch; in all instances, the actual final-crop trees will be fewer and farther apart. The marking of potential winners is quite optional, however, and has little point by mid-rotation when the crop trees are probably quite obvious. Marking them becomes unnecessary as one gains experience.

In choosing the potential 'winners', look for:

- A straight upright stem without forking or other defect below 7m; vigorous, probably a dominant
- A well-balanced crown with healthy, dense foliage and, in broadleaves especially, freedom from squirrel damage
- Few epicormic branches, especially in oak
- Good distribution (although this is less important than the quality).

These trees having been selected, the object in each thinning operation should be to remove trees which are competing with them or threatening them.

The trees to be cut in thinning should be marked before any cutting begins. They can be marked by slashing the bark, which is cheap but drastic; until you are experienced it is better to mark them with paint (which is slower but allows second thoughts) on opposite sides of the stem about a metre from the ground so that from all sides you can see what decisions you have already made. Whether or not you have marked potential crop trees, try to consider a group of trees together, as if the stand is made up of cells each of which will produce a potential crop tree. When you approach a cell, identify the likely 'winners' among the stems and remove

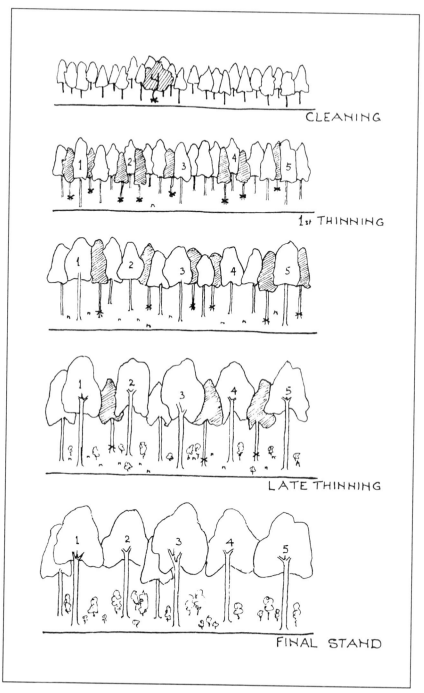

THINNING GUIDE

In cleaning (and perhaps in first thinning) concentrate on removing undesirable trees (wolves, misshapen stems, etc.). As early as possible identify likely 'winners' (well-shaped, straight, vigorous, well-spaced) and 'back' them by continually removing neighbours which are beginning to compete or threaten them.

trees competing with them or damaging them. The suppressed trees are irrelevant in this respect, although thinning beginners always concentrate on removing this class because it is the easy option. Indeed, the small trees near the winners may be beneficial by limiting the growth of side branches.

Do not attempt to rectify all the stand's faults in one visit. Too violent opening of a dense stand will do more harm than good, even leading to instability and destruction in the next gale or heavy snowfall. A useful guide to the intensity of thinning required is the depth of live crown on the pole-sized trees; in most species this should be about 40 per cent of the tree's total height, in some – larch, for instance – 50 per cent is better. The gaps in the canopy created at each thinning should be capable of being filled by the growth of neighbouring trees by the time the next thinning visit is due.

As a guide, the first thinning in a good young conifer stand may remove a third of the trees; the percentage removed would then drop steadily at each subsequent thinning down to 10 per cent in late ones. These are only guides, however; the criterion must be the improvement of the stand.

Delaying first thinning is particularly damaging in conifers, leading to unduly small crowns and stem weakness from which they recover only slowly, if at all. The root system can grow and the stem thicken only after the leaf-area has increased to produce extra energy, but the leaf-area can increase only if the root system is providing more nutrients. The cross-sectional area of the sapwood is strictly proportional to the leaf-area; heartwood, once formed, cannot be retrieved to conduct sap, so the limited sapwood restricts the ability of the tree to grow out of its small crown. (This 'chicken-or-egg' anomaly still baffles plant scientists.) Some species, for instance larches, are particularly difficult to recover from delayed thinning. In a crowded plantation larch will grow in height with an ever-diminishing crown percentage until the trees are like fishing rods with live branches on only 12 per cent of their height, and when thinned they may simply bend over; at best the trees will thresh each other in a wind and will probably never recover their

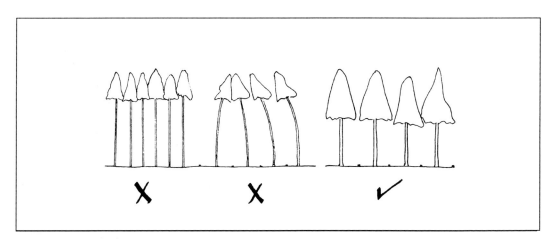

UNDERTHINNED LARCH

Larch stands in particular, but other species also, are seriously damaged by neglecting to thin at the due time or by underthinning. The congested crowns become small and the resultant 'fishing rods' cannot be relied on even to stand up, especially after a (too-late) thinning. Thin early and keep deep crowns on the trees.

vigour and growth potential. The only treatment for such a stand (apart from felling and starting afresh) is to make frequent, very light thinnings, say at three-year intervals and taking 10 per cent of the trees at each visit. Even that, however, may not succeed.

When thinning a mixed plantation containing a nurse, the forester must have the well-being of the main species firmly in mind. For instance, in a strip mixture of three rows pine, five rows oak, three rows pine, etc., which is growing well, it will almost always be necessary at first thinning to remove both outside rows of pine, or at least the whole row on the south side of the oak, in order to prevent the latter being suppressed. Similarly, in a mixture comprising broadleaved groups in a conifer matrix, the outside row of conifers will probably require removal at first thinning, as well as selective cutting in the interior. In all mixed plantings where conifers are used to nurse broadleaves, it is important to avoid suppressing and checking the vigorous growth of the latter; once the vigour has been checked, it is often difficult to recover the situation. This requires regular inspection and prompt response to a perceived need for opening.

Heavy thinning of pedunculate oak, for which there has been a vogue in recent years under the term 'free-growth', is associated with the formation of epicormic branches. These arise directly on the bole of the tree and, when they persist, cause a serious defect of the timber. Initially they form a thicket of short green shoots on the main stem and large branches; unless quickly killed by shading, some grow into substantial branches. The trigger for their development appears to be a change in the water relations of the trees, although some trees may be genetically prone to produce epicormic shoots.

The easy thing to do in thinning is to mark the trees which are already suppressed (even dead) whose removal will have no effect on the future of the main stand and will certainly bring in little or no revenue from sales. Leaving these trees may be helpful for wildlife, as are some small piles of brushwood, the lop and top of the cut trees.

Before any thinning operation is started, however, it is necessary to consider how the cut stems are to be removed from the wood. The smallest may be carried by hand but most will have to be drawn out by horse or tractor. The planning of the extraction system is important in order to avoid damaging the bases of the trees which are remaining and which are, therefore, the capital for the future.

The capacity of the woodland to operate as a sustainable resource depends on maintaining the cycling of nutrients and energy. That is why it is essential always to consider the ecosystem as a whole; anything which impairs these cycles – fire, soil compaction, loss of leaf-litter, too many deer, acidification, inept harvesting and other upsets – threatens the sustainability of the system and the future of the woodland.

NATURAL REGENERATION

In the interests of 'naturalness' there is often a desire to replace a mature stand of trees by natural regeneration. It is well to consider the constraints alongside the attractions.

- Natural regeneration depends on seed being produced from nearby trees. Most wind-dispersed seed falls within 100m of its parent, heavier seed much closer (except for the unreliable actions of crows), so vaguely hoping for trees to occupy a distant site is unrealistic. The parent trees must be of seed-bearing age and in vigorous condition to produce abundant seed, and they must be well distributed over the regeneration area, especially the heavy-seeded species such as oak and beech.
- Inevitably, natural regeneration repeats the

species and the genetic make-up of the parents. This may be an advantage if the parents are excellent but it is undesirable to allow bad parents to reproduce (poor form, low productivity, wrong species for the site, parents were foreign introductions, etc.). It may be quite naïve to think that the parent trees, because they are old, must be native; they may well be but adulteration of native woods has been active for at least 200 years, even in remote areas. In a derelict wood there may have been repeated removal of the best trees, the straightest and the most vigorous, so that the remainder – the potential seed-bearers – may be genetically inferior.

- Natural regeneration is not free, in the sense that it usually takes longer to achieve than establishing a new stand by planting and, in the extra time, heavier weed growth arises, thus increasing the tending costs. Even ignoring the delay in beginning the new rotation, the saving generally amounts to no more than the cost of the plants; tending costs may be similar, perhaps even greater.

- If successful, natural regeneration may provide a great benefit in the number of seedlings from which the new stand must be chosen, far greater than could be afforded in planting; in the rapid domination of the site; and in the desirable competition in the young stand, resulting in smaller branch development in the butt log and probably in a smaller proportion of juvenile wood. Some additional cost of the cleaning operation may result.

- Natural regeneration depends acutely on the state of the parent stand and, in particular, on the nature of the ground vegetation present in it. If there is a thick 'duff' of undecomposed litter, seeds may germinate but not survive (e.g. in a stand of pine, which is one reason for fire being

ecologically critical in natural pine forest). Seedlings of species with little stored food in the seed (birch, willow, poplar, etc.) must be able to take in moisture and nutrients immediately on germination; they cannot survive periods of adverse conditions or compete with other vegetation. The presence of heavy herbaceous weeds, a sward of grass or woodrush, heavy bracken or thick heather are strong counter-indications for attempting natural regeneration of any species. The importance of the soil condition is evident in the prolific natural seedlings often arising on the disturbed mineral soils along roads in forest areas.

- In contrast to the last point, a full canopy of vigorous parent trees, perhaps with an understorey of woody shrubs and sparse ground vegetation, provides good starting conditions. The soil surface may be prepared, perhaps simply by the activity of removing the understorey or by modest cultivation, and the canopy can then be opened in conjunction with the production of seed and as the seedlings develop. The control of the ground vegetation before seeding can be done chemically and mechanically, but it is worth considering if it might be done biologically, by cattle or by pigs.

- Tree seedlings are the preferred food of browsing animals, especially deer; for successful regeneration they must be reduced or excluded. Grey squirrels, crows, mice and voles are voracious eaters of tree seed.

- The cutting of the mature trees will very likely result in a rise of the water table; heavy soils may become so wet that seedlings will not grow. An early task in such areas is to clear the drains.

- If natural seeding is to develop into regeneration, work is needed continually to improve conditions for the young trees, perhaps opening the canopy further every two or three years. In natural forest, crops of seedlings appear repeatedly, decade after decade, only to die off ecologically unrequired because the canopy of the parents' crowns is vigorous and complete. In managed woodland, the art is to create conditions which ensure the survival and healthy growth of one such seedling crop when it is judged the parent trees are best due for replacement.

UNDERWOOD

A condition common to a high proportion of the older broadleaved woods in Britain is the absence of underwood, the layer of small trees and shrubs which should provide low shelter. Without it the woodlands are cold, draughty places, with nothing to stop the wind sweeping through the open boles. This shortcoming has usually arisen because of grazing and browsing by domestic stock and deer, but sometimes after a history of repeated ground fires; the ground vegetation is often grasses and woodrush.

The improvement of such woodland's value for wildlife, game and recreation calls for the establishment of an underwood of small trees and shrubs: hazel, holly, yew, beech, elder and other shade-tolerant species are candidates, according to the region and local site factors. The first steps are to close the woodland to domestic stock and to protect it against ground fires. Thereafter work should start at the woodland edges, to establish either a hedge or a shrub border which will begin to recreate the warm, sheltered habitats that woodland should provide. Small groups of the desired understorey species can then be planted in the interior of the wood.

FURTHER READING

Buckley, G.P. (ed.) (1992), *Ecology and Management of Coppice Woodlands*, Chapman and Hall, London

Edlin, H.L. (1949), *Woodland Crafts in Britain*, Batsford, London

Evans, J. (1984), *Silviculture of Broadleaved Woodland*, Forestry Commission Bulletin 62, HMSO, London

Evans, J. (1988), *Natural Regeneration of Broadleaves*, Forestry Commission Bulletin 78, HMSO, London

Harmer, R. (1995), *Management of Coppice Stools*, Research Information Note 259, Forestry Commission, Edinburgh

Harmer, R. & Kerr, G. (1995), *Natural Regeneration of Broadleaved Trees*, Research Information Note 275, Forestry Commission, Edinburgh

Hart, C. (1995), *Alternative Silvicultural Systems to Clear Cutting in Britain*, Forestry Commission Bulletin 115, HMSO, London

Kerr, G. and Evans, J. (1993), *Growing Broadleaves for Timber*, Forestry Commission Handbook 9, HMSO, London

Malcolm, D.C., Evans, J. & Edwards, P.N. (eds.) (1982), *Broadleaves in Britain*, Proceedings of Conference at University of Technology, Loughborough, July 1982, Institute of Chartered Foresters and Forestry Commission, Edinburgh

Conservation of Old Woodlands

7

Cut forests when it is a matter of urgency, you may, but it is time to stop destroying them. Every Russian forest is cracking under the axe; millions of trees are perishing; the abodes of birds and beasts are being ravaged; rivers are becoming shallow and drying up; wonderful lands are disappearing without leaving a trace.

To fell a thousand trees, to destroy them for the sake of two or three roubles, for women's rags, whims, luxury – to destroy them so that posterity should curse our savagery.

Anton Chekhov, 'Dr Astrov' in *Uncle Vanya*, 1900

After Dr Astrov's statement of the case there is really not much more to be said about the effects of unsustainable management of forest resources; he covers them all, just about.

Ancient semi-natural woodlands have special value for ecological science and conservation, and they call for special attention from their owners. But what attention should they receive? And what is good practice?

Although the two terms are often rolled into one as 'ancient and semi-natural', the concepts are distinct. Ancient woodlands can be traced as having been woodland continuously for several hundred years, usually since about 1600 in England and Wales and since 1750 in Scotland, when General Roy surveyed the first large-scale military maps which showed the woods in detail. Since these dates are about the time when regular forest planting began, the areas then known as woodlands were probably survivors of the country's natural forest or wild wood. Ancient woodland may now contain old trees or direct descendants of the native ones, but this need not be so; the native trees may have been cut and other species planted. The interest of ancient woodland is potentially in the woodland ecology, in the whole assemblage of other plants, fungi and insects, because these organisms may be direct descendants of the local wild-wood components, having lived on at the site even if the tree species have changed.

The interest in semi-natural woods lies in the present tree stands which are mainly of native species and have arisen spontaneously, not having been planted. In order to qualify, the trees must be locally native; as is made clear in chapter three, some species are native to only part of Britain and their range has been increased artificially, like beech north to Scotland and Scots pine south to England.

Woods which are both ancient and presently composed of locally native species are of most importance for conservation.

153

Many have been designated by the nature conservation agencies as Sites of Special Scientific Interest (SSSI) as a way of ensuring they are cared for. The best of them are the nearest we have to the original natural woodlands, and their special value lies in the fact that the trees and the other organisms have been naturally selected over many generations as being well suited to that locality's soil and climate. Biologists want to guard against losing them and against crossbreeding.

There is a balance to be struck between idealist conservation and managing for timber. 'Useful' plants and animals (so regarded for their economic value) are much less at risk of reduction or extinction than superficially 'useless' ones like mosses and lichens. There is always a temptation for people to 'improve' areas of woodland which are unproductive and therefore apparently 'useless', so the long-term conservation of these woods will always be at risk. It is arguable, therefore, that the conservation of ecologically valuable habitats may be enhanced if they can be sympathetically managed both to produce financially high-value material on a long rotation and, at the same time, to provide the necessary habitats for rare creatures. For instance, if the objective is to conserve oak woodland for its whole complex of higher plants, mosses, lichens, liverworts and insects, the chances of the ecosystem surviving to the twenty-third century appear to be better if the oaks are well-grown timber trees regarded as economically valuable and managed as a sustainable resource than if they are crooked and scrubby. An alliance between the owners, conservationists and the timber merchant who values this as a sustainable source of sawlogs may be more robust than conservationists alone fighting rearguard actions to protect what some people will always find difficult to appreciate. There is a special place for areas reserved as wild wood purely for nature, but the power of

sustainable management, which is the mark of the true forester, should not be underestimated as a force for conservation of an ecosystem.

Some broad principles apply to the handling of all ancient and semi-natural woods:

- If the wood has been designated a Site of Special Scientific Interest or a similar protected area, consult the local officer of the country's nature conservation agency before taking any action which might harm it, including a change of the management such as a change of grazing regime. Potentially damaging operations will have been specified in the SSSI designation notice but this may be superseded by a positive management agreement with the agency.
- Cutting trees generally requires a felling licence from the Forestry Authority's local officer; he may be able to help with advice about financial grants available.
- It usually pays to inform the local wildlife trust secretary to forestall rumours and to reassure the trust about intentions; trust members may be willing to help with conservation work within a well-devised plan.

In all semi-natural woodlands, the owner's policy, while taking full opportunity to harvest useful produce, should be to maintain and restore the natural diversity and maintain the genetic integrity of the populations of native species. Within these objectives it may be possible to extend the woodlands and improve their aesthetic or landscape value, but these are secondary to the ecological diversity and the genetic integrity. Such woods normally qualify for a special rate of management grant from the Forestry Authority if work is to be done to improve their value; it is a requirement of the grants that a simple plan of the management proposals be prepared. Such a plan is well worth while for these woods, not merely to obtain a grant but to promote continuity of purpose.

The maintenance of 'genetic integrity' is important. The oak, ash, pine and other native species grew here for scores of generations with only trivial interference by Man, during which individual plants unsuited to the local climate were weeded out by competition from well-suited ones. What grew here in 1500 or 1600, before large afforestation began, were strains carefully selected by our local conditions. Just as the manager of a pedigree herd would not allow casual crossing with unknown animals from elsewhere, so we want to avoid casual introductions of trees to the relict areas of ancient semi-natural woodland.

In the past many of these unwise introductions have been made, including Scots pine from Germany into the Highlands, oak from Romania, Hungary and other countries, hawthorn from the Netherlands and Italy, and so on. Some of these introductions were well-intentioned, when estate-owners last century saw wonderful pine somewhere on the Continent and aimed to raise the productivity of their plantations (often without success because of climate differences), but many have been totally unnecessary and can be attributed solely to the nursery trade's convenience and profit, and the unscrupulousness of major buyers, looking for cheap plants at any ecological cost. There is not, and never has been, any case for importing plants of common hawthorn from central and southern Europe other than nursery profit, but it has occurred on a large scale despite there being plenty of native seed. The scandal in 1980 of 'sessile oak' sold by German nurseries with EU certificates purporting to show that the seed came from named registered stands – whereas it had been imported as pig feed from Romania – led not only to millions of trees being burned but also to damage on an unknown scale by adulteration of the genetic base of oak across Europe. Maintenance of tree genetic integrity is a difficult battle but an important one.

The immediate revival of concern about our native woodlands and the spur to their active conservation came in 1981 from the first publication of Dr George Peterken's book *Woodland Conservation and Management*, which considers in detail both the characteristics of the various types in Britain and good practice in managing them.

Eight groups of semi-natural woods cover the major former wild-wood types of the country. The first four types occur in north and west Britain. Numbers five to seven occur mostly in the south and east. Number eight occurs in all regions.

1. Native pinewoods
2. Upland oakwoods
3. Upland birchwoods
4. Upland mixed ashwoods
5. Lowland acid oak and beechwoods
6. Lowland beech-ashwoods
7. Lowland mixed broadleaves
8. Wet woodlands

The characteristics of each of these woodland types are outlined on the following pages. Precise proposals for their protection and management must be tailored to the local conditions. For those who want to create a woodland of native species, these semi-natural woodland types are the essential basis for the design and selection of species.

NATIVE PINEWOODS

The major work by Professor H.M. Steven and A. Carlisle, *The Native Pinewoods of Scotland*, published in 1959, was crucially influential in focusing attention on the condition and importance of these relict semi-natural woodlands.

After the publication of Dr Peterken's book in 1981 and the 'Broadleaves in Britain' conference in 1982, there was sharply renewed interest in the Scottish native pinewoods, and the Forestry Commission's 1989 publication of the management guidelines for them set working practices for their conservation.

The reduction of the once-extensive Caledonian pinewoods to the present fragments which amount to only about 16,000ha did not happen in pre-history; much of the destruction is historically recorded after 1750 by a combination of felling for timber and overstocking with sheep and deer which prevented regeneration. Following gross exaggeration of the supply of pine timber and its sustainability, there was heavy felling around 1780 for markets which included the boring out of tree trunks for main water-supply pipes in London. There was further heavy cutting during the two World Wars, after which grazing intensified. Especially damaging has been browsing by red deer in the last 30 years, when their numbers have increased dramatically; many pinewoods have been favourite winter shelter for large herds of deer and, as a result, the ground plants have suffered severely, including the tree seedlings. In spite of the history of felling and overgrazing, the remnants of the pinewoods form one of the least disturbed and least modified forest types in Britain, which it is vital to sustain.

These remnants are genetically important as the north-western outlier of Scots pine's huge range which extends south to Spain and eastwards far into Siberia. Biochemical analysis of the terpenes in the resin has shown that this is not a single population over the Scottish Highlands but comprises seven sub-groups. The main differences are between the stands of the drier east and the wet west coast.

In Highland woodlands the Scots pine is commonly mixed with downy and silver birches, rowan, aspen, juniper and sometimes with common alder, bird cherry, sessile oak and willows. The whole range was ice-covered in the last glaciation, about 9,000 years ago, and the soils are acid podzols mostly formed from glacial outwash sands and gravels, in morainic knolls, drumlins and ridges, with wetter hollows between, especially in the west. Typically pine occupies the dry soils of the knolls, with broadleaves on the flushed soils lower on the slopes, and the wettest hollows left as open peaty mires. The common ground plants are heather and bell heather, blaeberry (bilberry) and fine grasses such as wavy hair-grass, almost always with the *Hylocomium* feather mosses on the drier ground and *Sphagnum* mosses in the wet hollows, with several other rare mosses and liverworts. There are also less common flowering plants, such as the wintergreens and orchids – creeping ladies' tresses and lesser twayblade – as well as specialist insects and distinct birds, including the capercaillie, Scottish crossbill, black grouse and crested tit. The red squirrel, pine marten and wildcat are also at home here.

In all the native pinewoods the policy must be to maintain the genetic integrity of the stands of native species and to maintain or restore the natural ecological diversity. In many, the diversity has been seriously reduced by overstocking with sheep and deer so that over large areas there are only old pine trees standing in a sea of heather. Light grazing is natural in these woods but it must be light enough to allow both the most palatable ground plants to exist and, essentially, enough seedlings of pine and broadleaves to survive and so ensure that the forest can be sustained for ever. If there are only old pine and a few browsed seedlings just showing among the heather, but no groups of poles or saplings, that woodland is in trouble and at risk of extinction when the old trees cease to bear seed.

Regeneration of the locally native pine is best achieved by natural seeding and the only actions required may be to prepare a good seedbed and prevent browsing. If non-native conifers have been planted in a native pinewood they should be cut out, certainly before they begin seeding. The natural pine forest normally regenerated after severe fires which burned off the deep heather and dry pine litter. Fire is acceptable for the same purpose now in managed woodland, provided it does not get out of hand. In small areas burning is best avoided and the soil should be prepared by patchy scarification, but not done so as to leave the soil loose and dusty. Draining should not be done; on the dry pine ground it is unnecessary and in the wet hollows it would destroy conditions for broadleaves and rare plants without improving things for the pines. With some soil scarification (or burning in tight control) and fencing to exclude grazers for some years, or rigorous reduction of deer numbers, the pinewoods can be gradually extended by natural seeding and even by planting. If planting is used, care must be taken to use only the local native pine as the seed source. For both natural regeneration and planting of Caledonian pine, the Forestry Authority pays grants at higher than normal rates. On these acid soils no weeding should be necessary and no fertiliser should be applied.

Where regeneration is undertaken, the work should be on a scale appropriate to the silvicultural needs of the main trees. Both Scots pine and the birches require full light if they are to thrive and the openings of the canopy should be bold, especially in the eastern Highlands. A strip 50m by 100m would not be too large; the planting of small patches and single trees is generally a mistake and a waste of money, especially if undertaken

with inadequate protection from grazing animals. It is much better to concentrate efforts on a larger patch which can be looked after and which will result in the creation of a significant area of young woodland habitat to serve all the native plants and animals.

Pine produces excellent timber, most home supplies now coming from the extensive plantations in Moray and the eastern Highlands. The natural pine woods should yield harvestable timber in the future, when they are again in good heart. At present the essential management is to ensure their restoration to a sustainable condition. It is right that some old trees should be left unharvested for the benefit of the birds and insects which are part of the ecosystem.

By far the most urgent need in the pinewoods is to reduce the presently crippling level of grazing and browsing by sheep and deer, which has reduced the Caledonian Forest almost to extinction.

UPLAND OR WESTERN OAKWOODS

Ancient and semi-natural woods of this type, amounting to 60,000 or 70,000ha, occur from Cornwall to Sutherland, with main areas in Snowdonia, Cumbria and Argyll, where there is the greatest concentration. They grow on acid leached brown-earth soils, which are often thin and rocky, in cool, wet and windy places; although listed as 'upland', they grow almost at sea level in the north. The oak may be mixed with downy and silver birch, hazel, rowan, willows, holly and hawthorn, with occasional wild apple, bird cherry and aspen. Bracken is common, but before it unfolds many of these woods on lower ground are glorious with spring flowers such as primroses, wood anemone and bluebells, growing with sweet vernal and other grasses. On steeper ground the vegetation may be dominated by blaeberry and fine grasses.

There may be rarities, too, including mosses, liverworts and lichens, rare butterflies and other insects and, in the wilder places, rare mammals, including the pine marten. Pied flycatchers and the greater spotted woodpecker are among the interesting birds.

The typical oak here is the sessile, but most of the woods were managed for tan bark and gaps were planted with pedunculate oak, the easier seed to collect and raise in nurseries, so both species are often present. The proportion of trees other than oak is probably now lower than originally as a result of management. Throughout their range the woods were used over several centuries to make charcoal for metal smelting – copper, tin, silver, iron – and in the eighteenth and nineteenth centuries they were used extensively as a source of tan bark. For both uses coppicing was practised (18- to 35-year cycles were common), but whereas most trees can be made into charcoal, only pole-sized oak trees were wanted for tan-bark production and other species were rigorously removed. As a result, many of the western oakwoods are unnaturally pure. The coppicing for charcoal and tanning stopped before the end of the nineteenth century and the existing trees are generally over 100 years old. Over this century the woods have commonly been grazed by sheep or by sheep and deer, and natural seedlings or young trees are rare. Where the woods were managed as coppice-with-standards, the old oak standards may be especially valuable as hosts for bats and for rare lichens on their bark.

Detailed proposals for managing these woodlands depend on local conditions but the following points generally apply.

The scope for renewed coppicing is restricted both economically and ecologically. Markets for oak poles and tan bark are limited; if they exist, nurture and satisfy them with regular assured supplies. Biologically, there are difficulties: oak coppices strongly to about age 50, but old trees put up weak stool shoots, and trying to coppice at 120 years may fail.

The simplest helpful treatment is to ensure periodic 'holidays' from grazing and browsing by sheep and deer, for the whole wood or parts of it in turn, for sufficient years to allow oak seedlings to establish and grow beyond the size at which they can be destroyed by animals. If oak regenerates satisfactorily, other plants will probably benefit also. This prescription means fencing out the animals; protecting individual oak seedlings with tubes does nothing for the other organisms and generally produces far too few oak saplings.

About one year in five, in the autumn of the year following a warm, sunny summer when the trees store extra energy, oak produce a heavy crop of seed – a mast year – when up to 800,000 acorns per hectare may fall, weighing nearly a tonne. Peak seedfall is in October (in the north) to mid-November. Natural regeneration requires a good seedbed, sufficient light and protection from browsing animals. Of the huge crop of acorns, the great proportion are taken by pigeons, rooks, squirrels and mice, but there should still be plenty for germination. For centuries oak woodlands were subject to the right of 'pannage' by herds of pigs which, while rooting around to eat the acorns and ground plants, churned up the soil and trampled in more than sufficient seed to form thickets of oak for the next generation.

Young sessile oak seedlings tolerate some shade for a little time but should not be expected to grow properly in shady conditions. Oak saplings need light, and a clear window to the open sky. In order to achieve good germination, some cultivation of the soil may be essential using a farm harrow or cultivator, a rotovator or forking by hand, best done in late autumn just after seedfall. Survival of the seedlings will depend on reducing heavy weed growth, such as greater woodrush and brambles, and on rigorous exclusion of deer and sheep.

If you intend to regenerate oak woodland by natural seeding, concentrate your efforts by choosing the area which will be the target for full establishment of a new stand over a 15-year period, since heavy seed will not come every year. In a large oakwood the regeneration block might be a sixth of the area and after 15 years another sixth will be tackled; if that policy were held to, the whole wood should be regenerated in a century. The implications of that timescale should be considered. It would require some of the present trees, say ex-coppice now 100 years old, to wait for 100 years before they were felled, giving them a potential age of about 200 years, which might be quite feasible and acceptable. This kind of judgement, using the local ages and estimates, has to be made for each woodland, large or small: 'How long will these trees live or continue to produce seed?' The answer determines how fast the regeneration should proceed.

After the regeneration area has been chosen, the sequence of actions might be along the following lines:

1. Fence the area so that it is deer-proof; this is essential because oak plants are the preferred food of deer and sheep. The maximum tolerable number of animals is about four red deer or eight roe deer per 100ha; if there are more, regeneration will fail. It is better to have none.

2. If there is a complete canopy of old trees, prepare for regeneration by modest thinning, cutting trees other than oak (they will seed back into the area later) and any badly shaped oak with poor crowns. In western oakwoods, cut pedunculate oak where there is a choice, since they were probably introduced. Aim to get seed from strong, well-shaped sessile oak.

3. Oak seedlings survive and grow best in gaps. Within the chosen regeneration block, choose the places where work is to begin and cut trees to form gaps (or work from existing ones). A good rule for sessile oak is that a minimum gap for regeneration initially should measure across about twice the height of the surrounding tall trees. If the trees are 20m tall, the 'window' should be at least 40m across from the edges of the crowns (not from stem to stem), and preferably elongated north-south for better lighting.

4. When heavy seeding is expected (look at the flowering and acorn development in summer), rip or crush brambles and other heavy weeds in August or September, or treat them with a herbicide.

5. At the time of the heavy seedfall, cultivate or harrow the area of the gap in order to bury some of the seed, mainly to hide them from the pigeons and rooks. Shallow burial of acorns is helpful, increasing survival and germination. Good natural seeding should provide at least four plants per square metre, 40,000 to the hectare. Lack of attention to heavy weeds, seed predation and exposure to deer and rabbit browsing account for the poor results often reported.

6. When the seedlings of the initial gap are two years old, the gap must be increased in size. It is fatal to the seedlings if they remain shaded (and the gap will decrease as the old trees' crowns grow); waiting for more seedlings to germinate will jeopardise the existing ones. If there are blanks in the young stand, they may be filled by planting.

7. The first gaps may be extended or others may be opened elsewhere in the regeneration block until all has been treated and the new stand established.

8. At the end of the period, say 20 years, the regeneration of the first block must be completed, by natural seeding, direct

sowing or planting, and work begun on the next. If this is not so, it means the wood is in danger of dying of old age, since the mature trees have not raised a next generation.

Leaving the woodland 'to take its chance' in the unnatural conditions of heavy grazing is not a sustainable option. Man has interfered in so many ways that we must be prepared to manage these woods positively in order to sustain them. There is a very real danger that, in the presence of browsing and grazing animals, the western oakwoods will degenerate into birch and rowan. Nor should sentimentality prevent the cutting of over-mature trees; the lack of thicket and pole-sized oak in these woods is critical.

Where planting is necessary, oak should be put at a maximum of 1.8m spacing. Nurse trees may be used, especially in frosty and exposed places, using alder, birch and rowan; conifer nurses should not be used in ancient and semi-natural woods (although good practice elsewhere) except in some west of Scotland oakwoods, where pine is natural. Birch is likely to be the naturally seeded nurse, although rather less satisfactory than alder because of its tendency to thrash the oak crowns in winds. In all ancient and semi-natural woods only local seed should be used. The European Directive on Forest Reproductive Materials requires that oak plants for forestry planting should be raised only from registered seed stands, of which there are very few of this type, but plantings of 1,000 trees or less for conservation purposes may be taken from unregistered sources. Since conservation is important in all ancient and semi-natural woodlands, whether or not they produce timber as well, this is an important derogation, and it is a great merit of group regeneration that suitable groups, planted at a maximum of 1.8m spacing, require fewer than 1,000 plants. You should, therefore, use local sessile oak seed.

Really abundant natural seeding can usually beat weed growth, provided that prior cultivation of the seedbed has knocked back the strong weeds. Planted trees, inevitably at wider spacing, are much more at risk. A granular herbicide may be required for grassy areas and is much more effective than cutting, which does nothing to relieve root competition and the interception of water. (See chapter five for information on herbicides and their safe use.) Bracken may be tackled with herbicide or by breaking the fronds when they are very young and brittle.

For later care of oak stands, the tendency in Britain is to over-thin oak, following good practice in conifers. Good practice is to keep oak thickets and pole stands rather dense, continually removing crooked and coarse 'wolf' trees of oak and trees such as willow and birch, which may over-top the oak and shade them out. There should be diversity, so a proportion of birch, rowan, holly, bird cherry and aspen should be retained, although it should be remembered that it is an oak woodland that is being regenerated. Thereafter progressive opening of the crowns will allow large trees which have already grown tall, straight stems to develop. Sudden and heavy opening of the stands at an early age causes the trees to grow heavy, low branches and wide crowns and, in effect, produces scrub.

Invasion by rhododendron is a severe problem in many western oakwoods. Preventing or stopping the invasion, however tedious, is far easier than eradicating the established weed. Its removal must be given priority, using chemicals, cutting and grubbing up plants; special grants may be available from the Forestry Authority for rhododendron eradication in ancient and semi-natural oakwoods. Where there is a heavy infestation, say 3m high and solid over several hectares, it may be unrealistic to attempt restoration directly to oak; it may be better to reduce the rhododendron and to establish a dense conifer crop for a short rotation, after which oak woodland may be replanted.

In these oak woodlands it is often the case that there are areas of rocky ground and steep slopes on which the harvesting of any timber would be difficult and uneconomic. As a minimum conservation practice these areas should be left undisturbed, so that oak may grow to old age, then die and rot unharvested, for the benefit of beetles, lichens, birds and other organisms.

UPLAND BIRCHWOODS

This title covers a variety of woodlands, some birchwoods in their own right and some derived from upland oak, pine or ash woods and capable of reverting if seed is available. They occur very widely but are most common in Scotland, on free-draining acidic soils, generally over 250m elevation. The best, perhaps derived from destroyed oakwoods, are on degraded brown-earth soils, but such areas have been greatly reduced in recent decades by heavy stocking with sheep and by conifer planting; extensive stands on infertile soils are more common.

The woods may be composed of almost pure birch, both downy and silver, or may be mixed with associates such as rowan, aspen, sessile oak, Scots pine, hazel, ash, alder, willows, bird cherry and others, according to their location and history. Birches have a large number of insect and fungal associates and their conservation interest may be high.

Birch is a relatively short-lived tree, perhaps 80 years, after which the woodland may begin to evolve to another type. If cut or burnt when young, birch will regrow as coppice, but its capacity to do so diminishes sharply with age and it does not survive

repeated cutting. As a result coppice management of upland birch was uncommon and the woods mostly grow as high forest. Formerly large birch were used for furniture-making but in recent decades its use as sawn timber has been uncommon in Britain because of the poor form of most trees. This is in sharp contrast to experience in Scandinavia, where birch logs are both sawn and cut as veneers for plywood manufacture. Interest in the timber potential of birchwoods in Scotland is increasing.

Birches are pioneer trees, adapted for colonising bare soil, rock slips, erosion scars or burned areas. The seeds are tiny, with small energy stores, so that they have poor capacity for thrusting their roots through turf or competing with dense weeds when they are germinating. In general, birches require small areas of bare soil on which to germinate; cattle hoof-prints or molehills do well.

Upland birchwoods are usually on well-drained soil, although there may be wet flushes and small mires in hollows. In the interests of conservation and habitat diversity (as well as saving money), these should not be drained.

Where regeneration of semi-natural birchwood is required, do not use coppicing but rely on natural seeding; birches are difficult to transplant as bare-root seedlings, so planting should be avoided. If plants are needed for special work, use small cell-grown stock. For natural seeding it is essential to have good seedbed conditions; if necessary make small screefings or cultivations of turf areas where seed may fall and germinate – an occasional spadeful turned over will suffice. Young birch are intolerant of shade, so regeneration clearings should be large, not less than 50m by 100m. Some seeds travel long distances on the wind but most fall within 100m, so if seeding is wanted on areas larger than one hectare, a scattering of strong seed trees should be left at about 30m spacing, or ten per hectare.

Seed falls from the catkins whenever it is ripe, so collection of good seed for spot-sowing is not easy – the seed is either unripe or gone. If spot-sowing is wanted some distance from a mother tree, you might try the old forester's tip of collecting small branches with nearly ripe catkins and pegging them down with forked sticks on a scarified patch of soil so that the seed may ripen in the catkin and fall naturally.

Birch grows fast when young and should not require weeding in semi-natural woods. In unusual situations of heavy weed growth, tramping back tall, soft weeds which are overtopping seedlings may give adequate help or, if environmentally acceptable, herbicides might be used, although protecting the seedlings is difficult and the practice is usually out of place in extensive upland birchwoods.

The principal help for natural regeneration and its care is likely to be a substantial reduction of sheep and deer stocking, the former by fencing and the latter by shooting.

If thinning is to be done, it should begin at an early age and should keep half the total height of the tree as live crown. In very dense stands, birches may grow like fishing rods with tiny crowns; there is then no point in trying to thin these, since the remaining trees will bend over in wind and snow, and their crowns will never recover.

In many Scottish situations where birch-woods are in the range of native pine, it should be quite acceptable to see them change to Scots pine-birch woods. The potential timber production of the pine may make the enterprise financially more attractive than pure birch and, provided care is taken, there should be no loss of conservation value. Pine of a local strain should be planted on the well-drained ground, without draining the hollows which will sustain birch and other species. Small,

open areas are ecologically essential features of these woods and they are now 'paid for' in the Forestry Authority's grants.

UPLAND MIXED ASHWOODS

In various parts of upland Great Britain and Ireland, where the underlying rocks are fertile – on limestones, basalts and andesite, for instance – the natural woodland is likely to be of mixed broadleaves, with ash the dominant tree. There are examples on the Mendip Hills in Somerset, the Pennines in Derbyshire, Yorkshire and Northumberland, above Morecambe Bay in Lancashire, on the Ochil and Pentland Hills and the Campsie Fells in Central Scotland, extensively in central and western Ireland and in Lorne, Argyll and on the limestone areas near Ullapool in Wester Ross. The soils are base-rich brown earths, sometimes shallow and sometimes little more than rocky screes. Within the natural range of the beech much of this woodland is a stage towards beechwood, but elsewhere it maintains itself as ashwood. The soil is apparently too base-rich for oaks, and these are usually absent or rare.

In the south the common trees with ash are small-leaved lime, gean, hazel, whitebeam, field maple, hawthorn, holly and yew, with beech and sycamore as invaders, but in some woods the ash may be almost pure. The ground vegetation is usually dog's mercury, with ivy, bramble, wood anemone and shrubs such as dogwood.

In the north and in Ireland the ash is accompanied by rowan, alder, downy birch, holly and hazel. On low ground and where the rainfall is high the mixture is mostly ash and alder with willows, whereas in drier places and at higher elevations it is ash and birch. Ash-alder occurs in the west of Ireland counties of Sligo, Clare and Galway. Ash-birch occurs in Cumbria, Inverness-shire and County Antrim.

Ashwoods are among the richest habitats for wildlife of all kinds in Britain. There are flowering plants in abundance, including some extremely rare ones. The golden-grey bark of the mature trees holds unusual lichens. The rich plant life supports many insect species, including some rare butterflies.

In many of the woods there has long been intense grazing since these fertile brown-earth soils offer extremely valuable animal feeding, especially in otherwise infertile landscapes such as Wester Ross, and it is inevitable they should have been 'hammered' by cattle and sheep, perhaps for centuries. Such grazing has altered the ground vegetation profoundly. In places where the grazing is less intense, for instance on steep slopes, ash colonises readily with its frequent and abundant seeding.

Ash coppices well but it is only in the south of England that upland ashwoods are coppiced. If the produce can be used or sold, coppicing should continue; the system provides an especially valuable succession of habitats for wildlife. Traditionally ash was cut on rotations which varied locally from about seven years to 35, depending on the market for the produce: rake handles and scythe handles (stails and snaiths) on seven years, cleft handles and hafts for hammers and axes on 25 to 30 years, gate rails and turnery wood up to 35 years and the cleft material for barrel hoops, baskets and so on as by-products from a variety of rotations. Deer browsing must be strictly controlled or absent from areas of young coppice ash for both commercial and conservation reasons.

Only occasionally do the ash in the uplands now produce valuable timber, although sheltered streamsides provide exceptions. Managed for timber, ash may be cut on a rotation of about 50 to 70 years. If it is to fetch a good price, the timber should be white or slightly pink, with wide, even growth rings; wood which is a yellow and patchy brown colour with irregular rings is regarded as defective, especially in resilience. The elasticity of ash wood, its attribute for hammer handles and sports goods, is said to diminish with height above the ground, so the butt log is the most valuable.

If upland ashwoods are to be managed as multi-purpose woods they can be regenerated by natural seeding. When the time comes for that, the light requirements of the young plants should be remembered: they benefit from woodland side shelter from the wind but will tolerate shade for only a year or two. In most circumstances group regeneration is ideal, the opening 25–40m across (the long axis, if there is one, north-south for good lighting). After a few years the smaller groups will require to be extended to allow the cone of saplings to thrive, since beyond early life ash is an intense light-demander. Fencing at the seedling to thicket stages will usually be essential. Planting may supplement the natural seeding; 2m spacing is appropriate. Competition from a grass sward is a problem in grazed ashwoods, so weeding may be necessary, preferably using a herbicide, for about three years.

After successful regeneration, the saplings and poles must be allowed to keep growing vigorously by respacing and thinning. In dense thickets, crooked and damaged stems may be cut back, and the aim should be to keep live crown on 40–50 per cent of the stem of the trees which are going to form the growing stand.

LOWLAND BEECH-ASH WOODS

Within the natural range of beech in England, on the fertile, base-rich soils of the limestones, ashwoods tend to evolve into beechwoods. These occur on the Chilterns, the Cotswolds, the North and South Downs, in the lower Wye valley and in Mid-Glamorgan.

On the limestone escarpments and the plateaux above them, beech grows tall and vigorously with groups of ash occurring sporadically as gaps form; its greater shade-tolerance allows beech to grow under the ash, from which it can later take over. Other trees present are whitebeam and yew, and sometimes gean, wild service tree and field maple, but beech forms the great bulk of the stands. On the driest slopes, especially on the fiercely draining south slopes on chalk, the beechwoods may have an understorey of yew; there are usually no ground plants because of the intense shade but dog's mercury, arum and violets may exist in open patches.

It is likely that past management for timber has increased the proportion of beech in these stands and has reduced their diversity. Beech is an excellent and versatile timber for furniture-making; selection for it by foresters has converted mixed woods into virtually pure beech, backed by its prolific seeding, the heavy shade it casts and its own shade-tolerance.

Beech coppices moderately well and was formerly so managed, especially in the Cotswolds, where the stands include ash, lime and hazel, but most woods are managed as uneven-aged high forest. It is well-suited to group regeneration with a gap size between 0.15 and 0.25ha, and natural seeding is successful in the space opened by even one or two mature trees. This pattern of opening results in structural diversity, but if species diversity is also required, rather larger groups will be necessary. Felling for natural regeneration can often be arranged to take advantage of patches of already existing seedlings or advance growth. The greatest problem is usually presented by the heavy growth of bramble which is common in these woods, and stern measures have to be taken against it, by mechanical stripping and by the use of herbicides. Shade is the best prevention.

At the sapling and pole stages of beech, great damage is done by grey squirrels, which kill large numbers of beech; these animals must be controlled, in the interests of both timber and wildlife conservation. Old pollarded trees have high landscape value.

Beech transplants well with small, stocky plants, to fill any blank areas in natural seeding. It is subject to the Forest Reproductive Materials regulations and care should be taken to use plants grown from approved seed sources.

Early tending of pole stands of beech should aim to remove coarse 'wolf' trees and maintain a reasonable proportion of ash and other associated species. Care should be taken not to thin so heavily that a dense growth of brambles is encouraged. Respacing may be required at an earlier age but the first normal thinning of beech is usually at about 30 years. The timber rotation is commonly about 100 years, although a few trees may be allowed to grow longer and even to remain uncut for the benefit of wildlife, especially hole-nesting birds.

In southern England the beechwoods have great landscape importance, exceeding their wildlife value. This is especially true of the 'hangers' on the chalk scarps, which are woods of great beauty and historical significance.

LOWLAND ACID BEECH AND OAKWOODS

In southern England and south-east Wales the semi-natural woodlands on acid soils are dominated by oak and beech, often with birch or sweet chestnut. The soils are acid brown earths or podzols formed on sands and gravels, or heavy clays, generally unsuited to arable farming and left for woodland and grazing. Many famous ancient forests are of this type: Epping Forest, Windsor Great Park and Forest, Sherwood Forest and the New Forest. They afford a great amount of public recreation and considerable care with communication is needed in order to carry public opinion with the necessary practice for their conservation. Although many of the largest woodlands are in public ownership, many smaller areas are privately owned or are in trusts; all are valuable and deserve care.

The species are usually pedunculate oak and beech, sometimes sessile oak, with birches and occasionally whitebeam, hornbeam, alder, buckthorn, yew, holly and rowan. Continental European pedunculate oak, non-native Scots pine, sweet chestnut and sycamore have been introduced, as has rhododendron. These woodlands have a long history of intense management for Man's purposes, probably 2,000 years. They have been hunting forests, then open to domestic grazing and periodically over-cut for timber to build houses or ships. Occasionally a worried John Evelyn or Samuel Pepys has stimulated the authorities to take action to conserve and regenerate them. All in all, it is difficult now to conceive what they may have been like originally, although pollen analysis suggests they may have contained much more small-leaved lime than they do now. As may be seen south of London, where heavy grazing has been stopped, dense birch woodland grows; later this may be invaded by oak. Oakwoods themselves tend to be invaded by the shade-tolerant beech, all dependent on seed being available and brought in, mainly by birds. The sweet chestnut was introduced as plantations. In these woodlands there is usually considerable diversity of habitats with areas of heathland nearby, patches of old coppiced oak and hornbeam, strips of alder and wetland. Where grazing was severe, however, the loss of the underwood and progressive soil degradation has reduced the woodland to scattered senile trees in a sea of heathland.

Silvicultural practice must vary with differences of composition. In areas dominated by beech, regeneration can be achieved by small group fellings similar to those described in the preceding section. In primarily oak areas group fellings for natural seeding should be considerably larger, say initially 0.25–0.5ha, depending on the size of the whole wood. Most of the recommendations in the earlier section on upland oakwoods apply here, including the need to give the young oak the increased light of the larger group size and particularly the need for protection against rabbits, deer and sheep.

In all areas some planting may be called for to enrich the natural seeding with associate species or to fill blank areas. Enrichment should be done within three years of felling for natural seeding. Concentration of effort in group regeneration is likely to be much more effective in the long term than the planting of individual trees scattered through an open stand.

In areas which are to be planted, the use of conifer nurses may be controversial. Many conservationists have been prejudiced against them, partly in fear that they might not be removed and the plantation would become a pure conifer one. On the other hand, it is known that many of the stands of excellent oak and beech now so much admired were raised with conifer nurses in the early nineteenth century and it is somewhat illogical to deny the practice now, except in specially designated conservation areas. Larch or Scots pine nurses may be doubly beneficial in raising the quality of the broadleaved trees and providing an early income for the owner who might otherwise wait 50 years for his first cash. An important consideration should be whether the end result, the woodland of native species, would be more successful with, or without, the use of nurses. In any event, the work should be carefully planned and recorded in a simple management plan endorsed by the Forestry Authority and, if appropriate, by the nature conservation agency English Nature. Such plans should clearly prescribe the removal of any nurse trees to give confidence about the end result.

LOWLAND MIXED BROADLEAVED WOODS AND WET WOODLANDS

Apart from the semi-natural woodlands already described, there remains a wide range of woods in the lowlands which deserve careful conservation. The species represented in them include the oak, ash, birch and beech already described, with limes, gean, hazel, hornbeam and, until a few years ago, many elms. Much of this woodland, especially in the south of England, was run as coppice or coppice-with-standards and exists as isolated woods of 10–30ha in farmland. They usually have an understorey of shrubs of hawthorn, blackthorn, spindle, dogwood and elder.

The fact that these woods often stand as islands of trees in a sea of arable farmland gives them special value for conservation, since they are almost the only elements of the landscape with any diversity from the farm crops and the only areas which are not severely disturbed several times each year.

The treatment of these woodlands cannot be prescribed in general terms. Each must be handled according to its composition, its condition and the possible market for its products. Well-managed coppice has high value as wildlife habitat because it maintains the short cycle of light and shade in which lowland wildlife thrives and it creates an abundance of subtle changes of habitat detail in short distances. But there is no point in reworking the coppice if there is no market for the produce, since the management would always be dependent on the goodwill of volunteers and would probably fail within a rotation. If a market can be found, say for firewood and field craft industry, the reintroduction of coppicing should be attempted.

If coppice working is not feasible, the old coppice areas may best be managed by 'storing' and conversion to high forest. For species which grow to full tree stature, this entails the reduction of the many coppice shoots to one per stool and the filling of gaps with new planting. Although the timber value of the present stand may be low, the quality of the next rotation can be high, with little or no reduction in the conservation value. Storing does not work with hazel; there is no alternative to replanting with another tree, although hazel might well remain as an understorey.

Regeneration may best be by a combination of natural seeding and planting. It may be possible to work the woodland in the same units as the former coppicing. In a small woodland this will usually mean there will be only a few age classes represented and it will be appropriate to space the regeneration fellings several years apart, to increase the age diversity. The spacing of the cuttings in time will depend in part on how long the last of the existing woodland can be expected to remain in place before it is regenerated, partly on the marketability of the present stand and partly on the size of the wood; there is no point in devising an elegant management programme if the working areas would be tiny and unworkable in practice.

Damage by grey squirrels is a major threat to the existence of these woods and control is essential. Deer, too, may be harmful, and their populations need management to allow the more palatable tree species to survive in the next generation.

Care should be taken of old pollarded trees, valuable as landscape features and as habitats for wild creatures, lichens, fungi and insects. Pollarding of new trees should be attempted only on the young to middle-aged, not on those already mature.

In several types of semi-natural woodland, the sycamore may present a problem. For 20 years or so conservationists were deeply opposed to it on the grounds it was an exotic which, at least in some situations, is strongly invasive and appeared to have fewer associated species – parasitic insects, fungi, etc. – than equivalent-stature native trees. On the other hand, sycamore is undoubtedly an excellent timber tree, capable of giving a considerably better yield than most native species. It is easily propagated, grows on a useful range of sites and withstands wind better than almost any other tree.

The conflict of conservation and timber interests with regard to sycamore has diminished recently; Peterken reappraises it in his 1993 edition, saying that sycamore has few damaging effects on ground vegetation and does support a rich store of epiphytes. He concludes that its usefulness outweighs its invasiveness, except that it must be eliminated from strict nature reserves and from those ancient woods where it is not already well established. This judgement is welcome since the tree, in many parts of the country, is not troublesome and, as 200-year stems, provides both focal points in the landscape and hosts for rare lichens, as well as being especially valuable for its timber and for its resistance to salt for coastal planting. Unfortunately it is exceedingly susceptible to damage by the grey squirrel.

Other special areas are the country's semi-natural wet woodlands, typically willow marshes, alder carr and wet birchwoods along the margins of lakes and mires and in the flood plains of rivers. Many of these have disappeared in recent decades, mainly as a result of district drainage schemes making wet woodland available for arable farming, and river-training projects, where planners and engineers thought – usually erroneously – they could avoid the inconvenience of rivers spilling out of their summer channels and thereby could allow houses to be built in the flood plains (with what appalling inconvenience periodically to their occupants).

Some alderwoods of this kind were formerly worked as coppice and some willow woods were pollarded, in both instances for the pole produce yielded. Many such woods, however, were not worked systematically; they might be cut irregularly, as produce was required by local people. The conservation value of the few remaining wet woods is now immeasurably higher than the produce they might provide and their management, if any, should reflect that.

Regeneration of willows is easy on this type of ground: strong, woody cuttings thrust into the ground generally root without problem. Birch and alder seed readily and, provided the water level allows it, natural seeding will recolonise sites. Many of these wet woods are designated Sites of Special Scientific Interest (SSSIs) and management proposals will have to be cleared with, or will be prescribed by, the nature conservation agency. The woods are usually even-aged, having regenerated after a catastrophe of some kind, perhaps excessive flooding or exploitation by local people. In such instances it is appropriate that future management should follow the same pattern. It would be a mistake, in a small wet wood, to attempt the creation of a regular series of age gradations by sustained patch cutting; better to protect the stand as long as it is healthy and then to ensure its complete replacement.

GENERAL

The problems associated with the management of ancient and semi-natural woodlands are complex and emotive. As a percentage of their original extent or of the total land area, they are minute and it may appear reasonable that they should be set aside for strict preservation. Nevertheless, their total area amounts to some 375,000ha and most are in private or trust ownership, so that the cost of even extensive care and the cost of foregoing their use are bound to be heavy.

Some of the woods deserve and require to be treated so that nature conservation is the only objective or consideration; all such areas, it is to be hoped, have been identified and designated as SSSIs, which will require the owner to consult the relevant nature conservation agency before undertaking work which might affect their well-being. They may well be subject to Management Agreements with the nature conservation agency, which will usually provide financial support for the prescribed management.

The bulk of the area, however, is not protected by formal designation and the challenge is to devise management regimes which will safeguard the ecological value while allowing compatible production of timber. In most instances this is fully possible, provided there is agreement to accept the constraints on the selection of tree species. In this regard it should be noted that the disparity in the productivity of trees between exotic conifers such as Douglas fir and native broadleaves is very large, probably eight times in terms of harvestable volume production per annum. The penalty attached to the species constraint appears to be severe, and it is one which should always be reflected in the grants of public money made available where woodland management is severely constrained by nature conservation; the size of the area to be conserved and the degree of constraint then become crucial issues.

In addressing the future management of ancient and semi-natural woodlands, much can be achieved by clear thinking about the objectives of conservation and the realities of timber harvesting. Clearly there is need for woodland owners to look realistically at the probable financial result of planting and tending proposed timber crops on some land; there are places from which no timber could be removed and even more on which no profit could be made by working timber. With benefit all round, such places can be left undisturbed or can be planted with native species without prospect of timber yield. On the other hand, the real requirements for effective conservation can easily be overstated: if the need is to sustain a particular native woodland ecosystem, it is likely it can be achieved at least as well by trees which are tall, straight and valuable as by those of the same species which are unmarketable and twisted. The high costs involved in maintaining large areas of native species with no intention of providing marketable material should not be disregarded, even if the whole cost is borne by the public purse.

Extreme positions are sometimes, but seldom, necessary or wise. Some sensitive woods must be protected with a view solely to their nature conservation value, but the natural heritage of many more can best be safeguarded by multi-purpose management which takes proper account of the non-market aspects as well as some timber production. In all there is need for the recording of intentions, actions and achievements in simple management plans which inform everyone concerned what is being done and why.

FURTHER READING

Aldhous, J.R. (ed.) (1994), *Our Pinewood Heritage*, Proceedings of Conference at Culloden Academy, Inverness, Forestry Commission, Royal Society for the Protection of Birds and Scottish Natural Heritage; Forestry Commission, Edinburgh

Anderson, M. L. (1967), *A History of Scottish Forestry*, Nelson, Edinburgh

Brown, I.R. (1983), *Management of Birch Woodland in Scotland*, Countryside Commission for Scotland, Perth

Buckley, G.P. (ed.) (1992), *Ecology and Management of Coppice Woodlands*, Chapman and Hall, London

Evans, J. (1984), *Silviculture of Broadleaved Woodland*, Forestry Commission Bulletin 62, HMSO, London

Forestry Commission (1989 *et seq.*), *The Management of Semi-Natural Woodlands*, Forestry Practice Guides 1 to 8, Forestry Commission, Edinburgh

Harding, P.T. and Rose, F. (1986), *Pasture Woodlands in Lowland Britain*, Institute of Terrestrial Ecology, Monk's Wood, Huntingdon

Kirby, B.J., Peterken, G.G., Spencer, J.W. and Walker, C.J. (1989), *Inventories of Ancient Semi-Natural Woodland*, Focus on Nature Conservation No. 6 (2nd ed.), Nature Conservancy Council, Peterborough

Linnard, W. (1982), *Welsh Woods and Forests: History and Utilisation*, National Museum of Wales

MacKenzie, N.A. & Callander, R.F. (1995), *The Native Woodland Resource in the Scottish Highlands*, Forestry Commission Technical Paper 12, Forestry Commission, Edinburgh

Malcolm, D.C., Evans, J. & Edwards, P.N. (eds.) (1982), *Broadleaves in Britain*, Proceedings of Conference at University of Technology, Loughborough, July 1982, Institute of Chartered Foresters and Forestry Commission, Edinburgh

Marran, P. (1992), *The Wild Woods. A Regional Guide to Britain's Ancient Woodland*, David and Charles, London

Miles, J. and Kinnaird, J.W. (1979), 'The Establishment and Regeneration of Birch, Juniper and Scots Pine in the Scottish Highlands', *Scottish Forestry* 33, 102–19

Patterson, G. (1993), *The Value of Birch in Upland Forests for Wildlife Conservation*, Forestry Commission Bulletin 109, HMSO, London

Peterken, G.F., (1993), *Woodland Conservation and Management* (2nd ed.), Chapman and Hall, London

Rodwell, J.S. (ed.) (1991), *British Plant Communities, Vol. 1. Woodlands and Scrub*, Cambridge University Press, Cambridge

Rodwell, J. and Patterson, G. (1994), *Creating New Native Woodlands*, Forestry Commission Bulletin 112, HMSO, London

Steven, M.M. and Carlisle, A. (1959), *The Native Pinewoods of Scotland*, Oliver and Boyd, Edinburgh

Worrell, R. (1996), *The Boreal Forests of Scotland*, Forestry Commission Technical Paper 14, Forestry Commission, Edinburgh

Protection of Trees and Woods 8

It is a basic reality of ecology that if an area (be it a forestry plantation or a wheatfield) offers a concentration of palatable and nutritious food, an organism of some kind – mammal, bird, insect, fungus, bacterium, virus – will find it advantageous to exploit that food supply. Their invasion is commonly considered harmful by mankind, especially by the person who has planted a crop, and protection is called for. Nevertheless, it must be recognised that the presence of wild creatures and their use of woodland resources is natural and often beneficial. They cannot all be excluded or avoided. What is generally more appropriate is to avoid a build-up of the invader's population to an epidemic.

The best protection for woodland in many instances is to practise good silviculture: to grow tree stands which are thoroughly suited to the locality, which are thriving and not physiologically stressed. The worst attacks are often associated with stands which are under stress, short of water, short of nutrients or under climatic stress. In such a state the stand not only is growing less vigorously but may have a reduced level of the natural chemical repellents which should give it protection from insects and grazers. Good selection of species and good silviculture are the first protective strategies.

BACTERIA AND FUNGI

In considering the relations between trees on the one hand and bacteria and fungi on the other, it must be remembered that the latter are not generally harmful but are very largely beneficial and strictly necessary for the well-being of the trees. Bacteria and fungi are essentially involved in breaking down leaf-litter for the recycling of nutrients and in the maintenance of soil structure. Both bacteria and some fungi even begin the breakdown process before the leaves of deciduous trees have fallen, germinating on the leaf surface in late summer and attacking the cuticle of the leaf while it is still functioning on the tree. A wide range of fungi, including many well-known toadstool species, are intimately associated with trees as mycorrhizas. These form a sheath of fungal threads around (and even into) the root tips of the trees, by means of which the tree can draw nourishment directly from the breakdown of organic litter. And some bacteria live in root nodules, supplying trees directly with nitrogen fixed from the air.

Hundreds of species of these groups of simple plants occur on trees and in woodland. Only seven are described here, those which may be particularly obvious to a visitor to woodland or which are nationally important in economic impact, or potentially so.

Bacteria are organisms without chlorophyll, single-celled or multi-celled, which increase by simple cell division. Fungi are multi-cellular, also without chlorophyll, relatively simple in structure, most reproducing by means of specialised spores released from mushrooms or similar structures.

Fire Fungus

The soil-living fungus *Rhizina undulata* causes root disease of young conifers. Spores in the soil germinate at a temperature of about 35°C, which is achieved only beneath a fire, and colonise the roots of newly cut conifers. From these they move to the roots of healthy young trees in an ever-widening circle (in France it is called *maladie du rond*). The fruiting bodies are bright chestnut-brown cushions on the soil surface, like a small heap of animal excrement, 4–12cm across. Almost all conifers are susceptible; Sitka spruce is especially prone, Douglas fir almost immune. Since it has ceased to be common practice in forestry to burn lop and top after felling, the disease is now rare, but owners of conifer woods should beware after fires. When an outbreak is young and the circle of infection small, control may be exercised by digging a trench around it, but this is hard work – and gets harder as the circle grows.

Fomes Root- and Butt-Rot

The Fomes fungus (*Heterobasidion annosum*) is the most serious disease of British forestry, potentially affecting almost all conifer stands. The fruiting bodies appear as narrow brackets at ground level on the stumps of conifers, red-brown in colour with white margins. Ideally the spores germinate on the newly cut surface of a conifer stump, for instance after a thinning; the fungus then colonises the stump, grows down through the whole root system and, where tree roots are in close contact, invades those of a neighbouring tree. The fungus then rots the roots of the standing tree and, depending on the species, may spread up the stem to rot that as well. Pines may be killed; spruces, larches and western hemlock suffer severe stem rot; Douglas fir and silver firs escape stem rot but may blow down after roots have rotted. Attacks are more acute on alkaline soils than on acid ones. This fungus is the only one in Britain against which protection is routinely practised; during thinning operations in conifer crops, immediately after each tree is cut, the stump should be painted with a chemical to kill the spores. Pine stumps, however, are better painted instead with a liquid suspension of the fungus *Peniophora gigantea*; this is another successful biological control developed by foresters, since the harmless *Peniophora* quickly occupies the stump surface and kills off any Fomes which arrive.

Honey Fungus

Honey fungus is, in fact, a group of several species of *Armillaria* all very similar in appearance and behaviour but each specialising in the tree species they attack. The occurrence is greater on newly replanted areas where old woodland has been cut down. The fruiting bodies are honey-scented toadstools which grow in abundant clumps around the bases of infected trees. The fungus spreads through the soil from infected stumps as long black 'shoelaces' which can then invade healthy trees, especially seedlings and saplings, which are killed. Affected trees will be found to have a sheet of fungus as thin as tissue-paper between the bark and the wood. Broadleaves, Douglas fir and silver firs are more resistant than the other conifers. The black 'shoelaces' of honey fungus are very common in woodland soils and stumps but they may be of a species which is not capable of invading living trees, so the forester should not jump to the conclusion that planted trees will die.

Dutch Elm Disease

This disease is caused by the fungus *Ophiostoma ulmi*, various strains of which have reached Britain over the past 60 years; the last import (on rock elm logs from America) was by far the most virulent and

HONEY FUNGUS
Armillaria species. *Honey-scented toadstools.*

devastating. The fungus grows from the cambium into the vessels in the sapwood and blocks them; as a result the twig or branch beyond the blockage point dies. The yellowing and withering of leaves well before autumn is a symptom. The disease is spread by elm bark beetles which feed in the branches early in the year before boring into the stem bark of a weakened tree. Sanitation felling to reduce the amount of infected material and the availability of good breeding trees for the beetles is the only counter-action. The beetle thrives only in mild climates; cold northern winters have helped to reduce the spread of the disease into north Scotland. The exotic English elm had virtually no genetic variation and none showed resistance to the disease; the native wych elm has much more genetic variation and shows more resistance, although certainly not immunity.

Ink Disease of Sweet Chestnut

This disease is caused by a soil-living species of *Phytophthora*; it causes sweet chestnut trees and coppice to die, after which it will be found that the affected roots and soil around them are stained inky black. The fungus spreads from the roots to kill spear-shaped patches of bark on the lower stem. The fungus requires free water in the soil for its development and neglect of drainage contributes to its spread. Control can be achieved only by drainage maintenance. After a severe attack, a change of species should be considered but beech should not be chosen as a replacement as it, too, is susceptible.

Until recently the *Phytophthora* species attacking sweet chestnut was the only one considered important in Britain, but recently the deaths of alders have been recorded in several river valleys in Britain, from the English Channel to the River Spey in Scotland. It is already clear that *Phytophthora* disease on alder is important; the symptoms are tarry or rusty spots on the bark of the main stem, near the base, coupled with mid-summer yellowing of the leaves which are often small, sparse and fall prematurely. Over a few years the tree dies. Observations of diseased alders should be reported to the research scientists in the Forestry Commission.

Pine Resin Top Disease

This is an endemic disease of Scots pine most common around the Moray Firth but occurring in several other low-rainfall areas, including East Anglia. It is caused by the parasitic rust fungus *Peridermium pini*. The trees attacked are usually large pole-sized, vigorous dominants which are potential final-crop trees. The fungus normally invades young twigs and grows down the branch to the main stem which it girdles, causing the whole crown above that point to die, usually with much exudation of resin. Masses of bright orange fruiting bodies are produced on

PINE RESIN TOP
Peridermium pini. *Dead tops of Scots pine, with bright orange fruiting pustules.*

the dead bark and a great abundance of spores are released in early summer. Unlike most rust fungi, this species has no alternate host, so the spores germinate directly on other Scots pine twigs. Locally this is a serious disease of Scots

pine, likely to increase as the Caledonian pine programme is advanced in Scotland. The only protection is to conduct rigorous sanitation fellings of affected trees and to destroy spore-bearing material.

Bacterial Canker of Poplar

The cankering and die-back of poplars, caused by *Xanthomonas populi*, is the most serious bacterial disease of trees in Britain. The cankers on the trunk make the wood useless for veneering. The bacteria are washed out of the cankers in rainwater and later, when dry, may be spread by the wind to invade other trees. Formerly, sanitation fellings were advocated, but the only effective action is to limit commercial planting to species and cultivars which are disease-resistant (see chapter three); planting grants are paid only for the approved varieties.

INSECTS

In spite of the huge number of insect species present in woodlands, there are very few which threaten the well-being of our trees. Those described here are grouped by lifestyle; the examples are chosen for their commonness and economic importance.

Defoliators

Defoliators eat the leaves of trees, their numbers sometimes exploding to epidemics. The presence of these insects can be easily detected by the loss of needles in conifers and of the leaves out of season in the broadleaves. As a result of trees losing their power to produce food, the timber growth slows, shown in the width of the annual growth rings; flowering and fruiting may be affected and the trees laid open to attack by other organisms. Serious defoliation for more than one year can kill conifers; broadleaves are generally more resilient and can survive several years' loss.

TORTRIX ON OAK
Tortrix viridana. *Early summer defoliation of oak.*

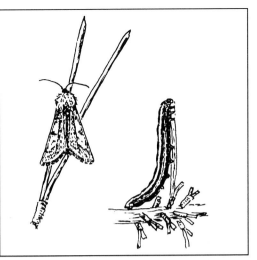

PINE BEAUTY MOTH
Panolis flammea. *Endemic on Scots pine but sometimes a killer of lodgepole pine.*

In natural forest, populations of defoliating insects may be cyclic; in one area the population will erupt, devastate the forest and then collapse under the combination of lack of food on the now-leafless trees and of the attacks of predators and parasites which move in on this abundant food supply of insects. In managed forest the owners are unwilling to allow things to take their course; they want the defoliator stopped long before the natural population crash, generally by spraying insecticide. This may be cost-effective in the short term but there may be adverse side-effects, such as the insecticide depleting the populations of 'good' insects, including the natural parasites of the target pest, and the increased risk of an insecticide-resistant strain of the pest developing. Intense entomological research into these matters goes on and recourse to the 'flit-gun' has been commendably restrained in British forestry.

Examples
The caterpillars of the oak leaf roller moth (*Tortrix viridana*) regularly strip the leaves from oak woodland in May or early June from the Midlands southwards. The greenish-grey larvae are quite small, only 12mm when full grown; if disturbed when feeding they quickly lower themselves on slender threads. Whole woods are completely defoliated, the trees forced to grow an entire new set of leaves, at high cost in timber growth. Pedunculate oak is the main species attacked; sessile oak flushes a few days later and largely escapes. The adult moth is green, the hind wings grey, total wingspan is 23mm. No protection is attempted; the side-effects of spraying would be unacceptable.

The pine beauty moth (*Panolis flammea*) is a natural associate of Scots pine, causing no serious harm in Britain although a real pest elsewhere in Europe. In his book *Forest Entomology* in 1908, Gillanders said, 'It is not often recorded as being numerically strong and therefore cannot be considered as a pest.' Without warning it emerged in 1976 as a serious defoliator of lodgepole pine, erupting as huge populations which have killed large areas of forest. This makes the point about the increased vulnerability of tree stands under stress; the pine plantations attacked were all on infertile peat, having just closed canopy and under severe physiological stress. As the Latin name suggests, the moth's forewings are bright flame-orange. There are few natural parasites at the outbreaks, and spraying has had to be used to control them.

The epidemic outbreaks of the pine saw-flies (*Diprion* and *Neodiprion spp*) and of the pine looper moth (*Bupalis piniaria*) have similar features to pine beauty moth.

Bark Beetles

Worldwide this group of beetles contains many hundreds of species, several of great economic importance, and the influence of a few has been considerable in Britain. As a rule, one beetle species is special to one tree species. The adult beetles excavate tunnels in the inner bark which carries the food from the leaves to the roots of the tree, laying rows of eggs along the sides of the tunnel. When the eggs hatch, dozens of grubs eat their way radially from the original egg gallery, mining the rich food supply. The pattern of their tunnels is characteristic for the species and their effect on the tree is fatal in many species, since the roots are starved of food and, with the root function impaired, the leaves go short of water. In many conifers there is an exudation of resin from the original entry hole and later there are very many exit holes of the next generation, after the fully grown grubs change into beetles. Bark beetles usually attack mature trees rather than saplings. Protection measures are difficult; the best are the use of traps baited with pheromones, the natural chemicals secreted by one sex of the insect to attract mates, and the breeding and release of numerous predatory insects which detect

and lay their eggs in the grubs of the bark beetles.

Examples

The great spruce bark beetle (*Dendroctonus micans*) first arrived in Britain in the early 1970s from Continental Europe (discovered only in 1982). It has established itself in most of Wales, the Welsh Marches and in Lancashire. In 1996 a further outbreak was found in Kent, probably with beetles imported in wood used as packaging material; woodland surveys led to sanitation fellings and restrictions on the movement of spruce logs and timber out of the area. Strenuous efforts are made to restrict the spread of the beetle, because potentially this is a tree-killer. The female, dull dark brown and 6–8mm long, bores into the spruce bark and lays eggs in a short tube. The larvae feed as a group, not in separate tunnels, excavating a large chamber in the bark. Abundant flows of resin on the outside of the bark are a feature. On

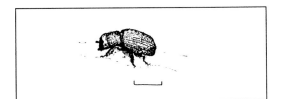

GREAT SPRUCE BARK BEETLE
Dendroctonus micans. *A serious recent introduction from Continental Europe.*

LARGE ELM BARK BEETLE
Scolytus scolytus. *One of several species responsible for spreading the Dutch elm disease fungus.*

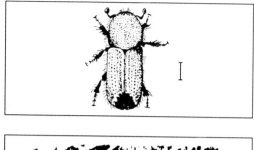

EIGHT-TOOTHED EUROPEAN SPRUCE
BARK BEETLE
Ips typographus. *The eight 'teeth' are spikes round its
tail-end. In a heavy attack every square centimeter of
the cambium is mined by the grubs, killing the tree
and leaving the wood surface under the bark 'engraved'
(hence the name* typographus*). It is 4–4.5mm long.*

the Continent it generally does not kill Norway spruce but whole woods of Sitka spruce have been killed. After careful research the Forestry Commission has developed good biological control with a predatory beetle, although, as is usual with predators, its ability to restrict the spread of low populations is questionable. This is a scheduled pest and all new findings of the insect outside its recorded occurrence (see above) must be reported to the Forestry Commission.

The large elm bark beetle (*Scolytus scolytus*) is one of several species which attack the elm. They feed on healthy twigs in the crowns of elms and then move to felled elm logs or to the boles of dying or weak elms for breeding. All these beetles may carry spores of the fungus causing Dutch elm disease. As the disease affects a tree, beginning with the infection in the branches, its weakness makes it ever more suitable for the bark beetle, so the population increases. The beetles' exit holes on the bole of the tree appear as shot holes. No effective control is known.

In the summer of 1997, during routine monitoring at Shotton paper mill near Liverpool, live adult eight-toothed spruce bark beetles (*Ips typographus*) were trapped. This beetle, hitherto unknown in Britain, is a serious killer of Norway spruce on Continental Europe; it can build up numbers on windblown or damaged trees before attacking nearby healthy forest which it can kill on a large scale. Whether the insects caught at Shotton came from a British forest or directly from imported timber (cargo packaging and dunnage are likely sources), it is important to maintain home woodlands free from this destructive pest. Any discovery of the eight-toothed spruce bark beetle should be reported immediately to an office of the Forestry Authority. Look out for it. Its arrival emphasises the importance of maintaining effective inspection at ports to intercept diseased material.

Bark Gnawers and Root Feeders

A group of beetles has evolved which specialise in attacking young tree seedlings, gnawing their thin bark and, very often, killing the plants by ring-barking them. Typically the beetles lay eggs in the stumps and roots of large trees. The emerging grubs burrow under the bark to feed and then emerge as adults to feed on the thin bark of the seedling regeneration, both conifers and broadleaves.

Examples

The pine weevil (*Hylobius abietis*) is one of the most troublesome of forest insect pests. The legless grubs, white with yellow heads, grow up to 2cm long in the stumps and roots of felled or dead conifers, living there for two years. The large adult weevil is black with a pattern of yellow scales, about 12mm long. The feeding adults (there may be as many as 150,000 per ha after felling a conifer crop) severely damage young plants and often kill them. Without adequate control measures, an average of 50 per cent of the young plants can be expected to be killed, such loss imposing quite unacceptable costs. The traditional protection, before the development of powerful insecticides, was to delay replanting felled conifers for three or four years, which allowed the weevil poulation to erupt and crash through lack of food, although the cost was high in extra unproductive time and in the heavy weed growth which developed in the interval. Foresters also did intensive trapping of the weevils and stripped the bark from stumps to destroy the breeding material; ineffective clearance of log off-cuts and poles after harvesting builds up weevil numbers. Recent practice has been to dip the plants in permerthrin insecticide immediately before planting and to reinforce that, if necessary, with subsequent spot-spraying on the young trees for one or two years. Close attention to health and safety measures are necessary with such treatment. An alternative to spot-spraying is to apply a 10g dose of carbosulfan granules to each hole at the time of planting; this is a slow-release systematic insecticide which should give at least two years' protection to the plant, with the advantage and safeguard that the only insects to be poisoned are those which begin to eat the bark. Even so, this is a serious and costly pest in all conifer areas and intense research is being conducted on biological controls using indigenous nematodes and parasitic Braconic wasps as natural control organisms which are environmentally safe and do not affect mammals, fish or birds. The development of such integrated pest-management systems may soon offer an effective control of pine weevil and thus avoid the reliance on insecticides. Biocontrol is the preferred forestry method of pest control.

The cockchafer (*Melalontha melalontha*) is one of several species which feed principally on the roots of young trees, especially in nurseries in southern England. The curved white grubs with brown heads live below ground and eat through the roots, causing the plants to wilt and die. The adults are big clumsy chafers, 30mm long, dull brown with a black head, which feed on the leaves of broadleaves. In northern England and Scotland a smaller insect, the brown chafer (*Serica brunnea*) lives in much the same way.

Sucking Insects

This group of insects has mouth parts specially adapted for sucking plant juices; they are aphids and include the ubiquitous green-fly. Their effect is to rob the tree of its food and water, perhaps causing the leaf which they

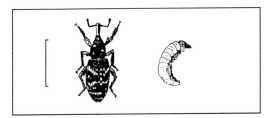

PINE WEEVIL
Hylobius abietis. A costly pest of conifer regeneration areas.

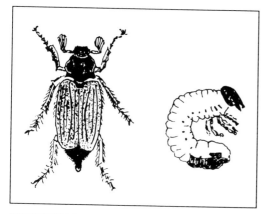

COCKCHAFER
Melalontha melalontha. *A pest of nursery seedbeds.*

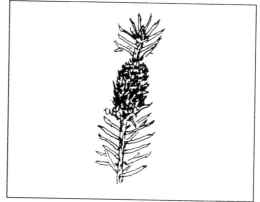

PINEAPPLE GALL ON SPRUCE
Adelges abietis. *Woody galls on Norway spruce twigs.*

have attacked to wither; reduced growth, impaired health or death of the tree may result.

Examples
The green spruce aphid (*Elatobium abietinum*) attacks spruces, especially Sitka spruce, feeding on the underside of the needles and causing them to fall off. In southern England the activity is almost confined to early summer, whereas in Ireland and Scotland attacks may be spread from spring to late autumn. The insect thrives in mild weather and hard frost is the main natural control of populations. Attacks are much more serious when the trees are stressed by moisture shortage and in low-rainfall districts whole woods of Sitka spruce may be virtually leafless in the winter after a severe attack, with great reduction of growth. The general rule is sound: Sitka spruce should not be planted in districts with less than 40 inches of rainfall (1,000mm). Even so, the insect will multiply after mild winters. Foresters never attempt control by insecticides, although Christmas-tree growers (on Norway spruce, etc.) and nurserymen certainly do.

The felted beech coccus (*Cryptococcus*

fagisuga) is a scale insect which forms large colonies on the bark of pole to mature beech trees, feeding through the bark of the stem and branches. The female insect itself is pale yellow, 1mm across, wingless and legless, covered with abundant white waxy felt. The attack may involve scattered groups of a few coccus, which do little damage, but it may extend unbroken over the whole stem of a large beech, the mass of adults, dead larvae and waxy felt 5 or 10mm thick, where large areas of bark are killed and separate from the stem. Individual specimen trees may be scrubbed and treated with a mild insecticide mixed with a wetting agent (soap or detergent) to 'cut' the waxy protection, but this is impossible on a forest scale. It may be simplistic to blame the death of large beech on this insect alone; heavy attacks are usually associated with water stress (after one or two dry years), or interference with rooting (beech feeding roots may be only a centimetre or two below the soil surface), or urban pollution, etc. Curiously, copper beech are much less prone to attack than common beech.

The pineapple gall aphid (*Adelges abietis*) is typical of another group of sucking insects. The adult female lays eggs below a bud on a

179

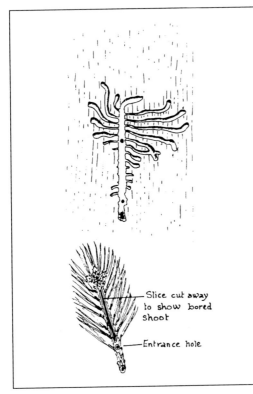

PINE SHOOT BORER
Tomicus piniperda. *Look for fallen green shoots under Scots pine.*

shoot of Norway spruce, protecting them with a covering of white cottony fibre. As the shoot lengthens, the irritation of the feeding larvae causes the bases of the young leaves to grow into a swollen gall, like a miniature pineapple. This later grows woody and black with slits which allow the larvae to emerge. A gall, about 5cm long, may contain many hundred larvae. The shoot above the gall may continue to grow or it may die; the condition is very disfiguring. This is a serious pest of Christmas-tree farms, controlled by contact insecticides. Many related species attack Douglas fir and other conifer trees; most occur on the foliage, with waxy white protective coats. No counter-measures are usually attempted in forest conditions.

Shoot Borers

These insects feed, and live for at least part of their life cycle, on the young twigs of trees, generally in the area immediately below the terminal bud where there is a concentration of food in the growing season.

Example

The pine shoot borer (*Tomicus piniperda*) is typical of this group. Evidence of its work is easy to find in Scots pine forest and plantations. It is a bark beetle which, within its life cycle, also bores into pine twigs. The population 'ticks over' at a moderate level, ready to explode in any situation which causes trees to be stressed or damaged. In spring, female beetles bore into the bark of pine logs or the bole of a sickly tree, excavating first a breeding chamber and then a 10cm-long straight mother tunnel. From this the larvae eat their feeding galleries, at first narrow, later widening as they grow, with adjacent colonies merging. During the summer these new adults and those of a second brood fly to the upper crowns of the Scots pine, where they bore into leading shoots about 10cm behind the bud and tunnel into the pith. The boring so weakens the twig that it snaps off in a wind; the ground in a pine plantation may be strewn with scores of these twigs, each bored up the pith and containing a bronze-coloured adult beetle about 5mm long. In stands of pole-sized Scots pine affected by pine shoot borer, the crowns are curiously pruned, having been flat-topped over years when the beetle population was high, then growing fast in narrow spires when the shoots were relieved of their attacks. Control is best exercised by good silviculture and promptly clearing logs which are potential breeding material.

WITCH'S BROOMS

A witch's broom is an abnormal dense tuft or mass of twigs growing on a branch. It is not strictly a subject for protection, but this is a convenient point at which to describe them. Hundreds of closely packed buds are stimulated to grow only as short shoots, probably always through a local upset in the tree's growth-control chemicals arising from an insect attack or a viral infection. Small brooms, say 30cm across, are quite common locally on downy birch, less frequent on other broadleaves; large brooms, up to 3m across, are seen occasionally on Scots pine or Norway spruce.

MAMMALS

Domestic stock, cattle, horses, sheep, goats, etc., damage woodland by soil compaction through trampling, browsing seedlings and, horses at least, gnawing the bark from even mature trees. Some farmers will argue for allowing animals to shelter in woodland in winter to avoid the wind and radiation frost and to be kept convenient for hand-feeding. Nonetheless it should be recognised that the heavy stocking (as this use always is), even for a week or two, imposes a high cost in damage by soil compaction and intense fouling of the ground with urine and dung. As a rule, livestock should be fenced out of small woodlands and, on larger areas, allowed to graze only occasionally and in small numbers.

Wild animals can have a dramatic effect on both planted and natural regeneration in both broadleaved and conifer woods; tree seedlings are high on the palatability ratings of most animals living in and near the woodlands. Control of herbivore animals is essential for the successful establishment and regeneration of woodlands. Which species are important is a local matter; how much should be spent on protection depends on the severity of the threat and what is at risk. The degree of protection provided must be balanced with the value of the trees at risk. It is worth spending money and effort up to the point when the cost of the last unit of protection (shooting, fencing, spraying, etc.) equals the value of the damage actually saved by it – and not a penny more.

Protection from animal damage can be achieved by:

- reducing the population so that damage is acceptable;
- fencing the animals out;
- protecting individual trees with tubes or repellents.

Field Voles

The field vole, *Microtis agrestis*, lives and feeds in grassy areas. The closure of pasture to sheep-grazing for tree-planting suits the vole well and the local population thrives. The populations tend to fluctuate naturally on a three- to five-year cycle, and when in numbers they gnaw the bark off young trees at the root collar, killing them by ring-barking and even felling them. Numbers are so great that trapping would be hopeless. The first protection is to avoid planting trees in a grass sward; a bare-earth zone around each tree not only relieves plant competition but means that voles would have to sit in the open to eat the tree instead of doing so under grass cover, thus exposing themselves to their natural predators. As the vole population increases locally, so the predators do better; owls, kestrels, weasels and foxes all adjust their breeding success to the availability of voles. All these predators do a good job.

Plastic vole guards are available, but are justified only if the population is expected to increase dramatically. As soon as the trees are dominating the grass they should be safe because the ideal vole habitat will have gone.

Rabbits

These are extremely damaging to woodlands, eating young plants to the ground and in winter gnawing the bark off saplings, particularly in snowy periods when other food may be covered. Rough-barked trees like Scots pine may become safe after about ten years but smooth-barked ones like ash and sycamore may suffer bark damage until 20 years or more. If rabbits are common it is a waste of money to plant anything until they have been killed or fenced out. (Landholders have a legal responsibility to control rabbits; success always depends on neighbourly collaboration.) Winter poisoning with cyanide in burrows, backed by intensive shooting of those living above ground, is usually the best approach.

Rabbit fencing should be of hexagonal mesh netting, 18-gauge mild steel galvanised wire, the maximum mesh size 31mm; when erected it should stand at least 900mm high with 150mm turned outward on the ground surface, pegged and turfed down (ie. the roll 1,050mm wide). As an alternative to fencing, treeshelters may be considered, but attention must be paid to the period of risk which may be well beyond the life of the tubes; unlike voles, rabbits are not deterred by the trees closing canopy.

Hares

Both the brown hare in the lowlands and the blue hare in the mountains can be troublesome in the year of planting and perhaps the next, while the area is still open for running. The damage is usually infuriating rather than disastrous; in winter and spring hares move down a row of trees and cleanly slice off each plant so that it lies neatly, uneaten, beside its stump. Broadleaves at least may sprout again from the base. Hares may be able to jump over minimum-height rabbit fences. The best plant protection is to shoot the hare.

Deer

Red, sika, roe, fallow and muntjac deer all live in woodland and have trees high on their preferred food rating. In addition to eating them, deer use trees to mark territories (roe bucks rub scent glands on the stems), to rub off the withering velvet from their new antlers, to thresh (perhaps just temper or an equivalent of shadow-boxing?) and, red deer at least, to strip bark for eating in winter.

Mankind has removed all effective predators of deer from Britain, so the only protections are shooting, fencing and treeshelters. All deer species except muntjac have close seasons for shooting, although permission may be sought to shoot those doing substantial damage to plantations in the close season.

The difficulty of controlling deer numbers in woodland is considerable. They are secretive (sika exceptionally so) and try to stay in thick cover in daylight. All respond to the sheltered conditions and good feeding which woodlands provide by maturing earlier, breeding faster and surviving better than in open habitats. Experience suggests that the safe stock of deer to allow natural regeneation of palatable species is about four red or sika deer per 100ha, or six fallow, or eight roe deer. Heavier stocking means that regeneration is progressively restricted to the less palatable species, losing the oak, beech, silver fir, etc., as well as sensitive rare plants which are the concern of the nature conservation agencies. The most unpalatable seedlings are Sitka spruce, so tolerating high deer populations means that this tree is favoured at the expense of other species.

Where a modest deer population lives in woodland, fair protection can be given to the leading shoots of selected seedlings, especially conifers, by wrapping a wisp of sheep's wool around the terminal bud; it seems deer are as sensitive as humans to the feeling of strands of wool on their lips, and they leave these shoots unharmed. In sheep country there is always some wool available on the fence wires for this low-cost protection.

In small woodlands, shooting from high seats is safest and most effective. It is best done in the very early morning when there are no people about, but it presents special problems where woods are well used by people. The compensation is that woods used by people, especially with dogs, are made unattractive for deer and the need for culling may be much reduced.

Grey Squirrel

This foreign import has unfortunately been so successful in displacing the native red squirrel that over most of Britain people know only this pest and think it is nice to see. It eats seeds, cones and berries, as well as buds and young twigs, which is all tolerable, but in addition it causes severe damage, both ecologically and economically, by stripping the bark from pole-sized tree stems. The stripping is usually done in high summer and trees are often ring-barked and killed; most are thin-barked broadleaves, beech and sycamore probably the worst affected.

In most of England and Wales, and in some districts of Scotland where red squirrels are not present, the best control is achieved by poisoning with 0.2 per cent warfarin on whole wheat, supplied in special tunnel dispensers which are designed to allow only grey squirrels to enter. Elsewhere, including most of Scotland, where poisoning is not permitted, cage-trapping followed by humane killing may be practised or the nests or drays may be destroyed by poking them with a pole and the squirrels shot. If high-quality hardwoods of the thin-barked species are to be grown, this destructive animal must be locally eliminated.

The red squirrel is a protected species in the schedules of the Wildlife and Countryside Act; its numbers are much smaller than those of the grey and it is generally easily accepted in high-quality woodlands. Nevertheless, at high concentrations it can do great damage, especially to Scots pine, by stripping the bark on the upper bole. The population explosion in parts of Scotland in about 1900 caused much loss over a large area. Where local damage is severe, permission may be sought from the nature conservation agency for live trapping and for the surplus animals to be sent to a suitable forest which has no squirrels. The upsurge in planting of Caledonian pine in Scotland may trigger a substantial increase in red squirrel, and some increase in damage should be expected.

CHEMICAL REPELLENTS

Chemical repellents are potentially an alternative to fencing and treeshelters to protect trees from browsing animals. Many are marketed but few are effective.

Tests by the Research Division of the Forestry Commission have shown one chemical to be effective when applied to the whole tree. *Aaprotect* has ziram as its active ingredient and it repels browsers and bark-gnawing mammals, from deer to voles, by irritating the animal's nose and mouth. Diluted 1:1 with water, it may be applied to the tree as a low-volume spray or may be painted or smeared on, which is less wasteful on small plants. Only the parts of the tree actually treated with the repellent are protected, so new spring growth is vulnerable; the chemical must be applied and reapplied annually, between mid-November and the end of February, as long as the tree is vulnerable. New planting may be treated with a halfstrength solution (i.e. 1:2 water). Protection of individual trees is more likely to be cost-effective than treating large areas of planting. Costs for the material only may range from £0.10 to £0.50 per tree per annum, depending on the size of the plant, or on an area basis, including labour costs, from £300 to £1,200 per ha per annum.

Since *Aaprotect* is an irritant to eyes, skin, nose and lungs, appropriate protective clothing must be worn during mixing and application (including rubber boots, rubber gloves, face shield and cover-all), and those who use it should be trained and certificated.

FENCING OR TREESHELTERS?

For protection against animal damage there may be a choice between area-fencing and individual tree protection (treeshelters or netting guards). The decision depends on:
- Size and shape of the area to be protected, affecting the length of the fence required;
- Number of trees to be protected;
- The kinds of animals involved, affecting the height of the fence and the treeshelters;
- Aesthetics, including people's freedom of movement;
- The risk of vandalism, since tubes make trees obvious and attract vandals in urban fringes.

Treeshelters are translucent polypropylene tubes which both protect the tree from animal browsing and give it a mini-greenhouse, so their use may increase height growth (but not stem thickness or root growth) due to higher air humidity and temperature. A few species respond very well in height growth, oak and sweet chestnut especially; others benefit little and some scarcely need help with height growth, e.g. ash, birch and sycamore; a few may be retarded, e.g. lime. Few conifers do well in shelters.

Treeshelters require stakes. A 25mm x 25mm stake is adequate for a 1.2m tube on a sheltered site, 30mm x 30mm on an exposed one, always shorter than the tube and firmly tied to it. A lighter stake which rots and snaps causes the tube to fall and ruin the tree. On a windy site, treeshelters are very troublesome and are best avoided; a leaning tube produces a leaning, bent tree.

The treeshelter must be above the browse level of the animals present. They will not protect against cattle or horses. Some breeds of sheep will climb on the stakes, so tall shelters are needed, and, since sheep rub on them, double staking is advised, making this an expensive protection system (about £2.60 per tree, 1996 prices). Shelters of 60cm suffice for rabbits alone and may be unavoidable in small woods if neighbours refuse to control rabbits.

Several shapes of shelter are available. Square sections store flat and are easier to tie to the stake, but they may blow flat in high winds. Round section tubes do not distort in wind but may twist on the stake and then blow down. The top lip is best curved to avoid cutting the plant when it emerges. Tubes are intended to degrade in sunlight by year five, although they do not appear to be as biodegradable as predicted and clearing the pieces is essential if the countryside is not to be a clutter of plastic rubbish. The tube colour should blend with its surroundings and pale brown is probably best; white tubes look like tombstones, and some colours, such as bright blue, are eyesores and should be avoided.

A treeshelter is no substitute for weeding, although often weeding is neglected. It does protect the plant from herbicide sprays, thus reducing the cost of its application, but this is no advantage if there are weeds, equally protected, growing inside the tube. If a treeshelter is to be justified, the weeding must be effective, so that the tree's growth pays for the tube.

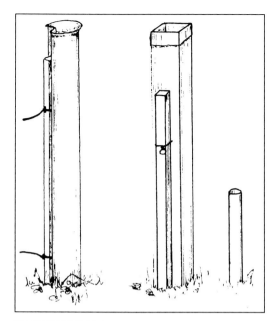

TREESHELTERS
Two types of treeshelter and (right) a 'quill'.

Trees growing in a tube have no stimulus to grow in thickness, so the plant, when it emerges, is tall and weak, unable to hold itself upright. This effect may be so severe that the forester may be forced to remove the tubes, cut the trees back to ground level (especially oak) and induce a vigorous shoot to grow from the base.

Treeshelters are very much in vogue and in some instances appear to have become almost a substitute for planting, showing a good intention even though there is nothing in the tubes except weeds.

The main problem is that treeshelters induce the planter to put in too few trees. The Forestry Authority in recent years has insisted that the *minimum* number of established trees should be 1,100 per ha, a spacing of 3m x 3m, a rule recently revoked. This is far too wide spacing for growing high-quality broadleaves; the minimum number per hectare should be 2,500 for most species, and 3,100 or more for oak and beech. But

treeshelters are costly, with stakes and fixing, and there is a temptation for those who have no vision of the future or who do not care about quality to plant the minimum number of plastic tubes to get the grant. The alternative is to fence and shoot the rabbits and control deer numbers.

Table 8.1

FENCING v. TREESHELTERS: PRIMARY COST
COMPARISONS
(excluding maintenance and dismantling)

Example 1
Rectangular site 200m x 30m; perimeter 460m;
Area 0.6 ha; number of plants 1,500; rabbits

Fencing	*Tubes*
460m @ £2.40 per m	1,500 @ £0.60
	(60cm shelters)
£1,104	£900 BEST BUY

Example 2
Rectangular site 200m x 100m; perimeter 600m;
Area 2.0 ha; number of plants 5,000; roe deer & rabbits

Fencing	*Tubes*
600m @ £4 per m	5,000 @ £1.10
	(1.2m shelters)
£2,400 BEST BUY	£5,500

In very small areas with few trees, tubes are usually less costly, but as the area increases in size, fencing becomes the better value, especially as many small woods and farm areas are already fenced. In making the comparison, remember the possibility of fencing the main area economically and leaving any awkward extensions for individual tree protection.

Plastic mesh guards are staked tubes of strong netting used to protect the main shoot of newly planted trees from deer damage by browsing and bark-stripping. They provide little shelter and are visually less intrusive than plastic tubes. Spiral guards are loosely coiled plastic strips used to protect against rabbit damage; they are suitable only for sturdy trees above 60cm high which can hold up the spiral unstaked. In contrast to treeshelters, spiral guards do allow flexing of the tree stem and therefore do not inhibit diameter growth.

Table 8.2

HEIGHTS FOR FENCES AND TREESHELTERS

Animals	Fences	Treeshelters
Voles	n/a	Vole guards 20cm buried >5mm in soil
Rabbits	Netting 90cm + 15cm at base turned out on soil surface and secured down	60cm
Hares	1m: as for rabbits + a wire 10cm above	75cm
Sheep	Standard stock fence	1.8m but double heavy stakes needed
Deer red and sika	2.0m	1.8m
roe and muntjac	1.8m	1.2m

Miniature treeshelters, called quills, are cylindrical plastic tubes 45 to 55cm tall by about 5cm in diameter, cut at an angle at the lower end. They can be used for hedge-planting and small trees, giving good protection against voles and some protection against rabbits and against herbicide spray. They do not require staking, the base of the narrow tube being buried in the soil. Quills should never be used with bare-rooted plants; thrusting them into the ground would slice off the roots. A cell-grown tree with roots in a 'plug' can be planted inside its quill without apparently suffering any root damage, although the effect on the subsequent growth of roots is unclear, since they may be prevented from spreading normally by the plastic. The containment of the shoot is a problem in hedge-planting, since pruning to induce early forking below 50cm height is most desirable. Following criticism that fast-growing trees were strangling in the tubes, which were not degrading sufficiently fast, manufacturers have made a line of fine perforations along the length of the tubes; one is left to clear up the plastic.

There may be occasions when treeshelters and quills are useful, but if animals can be kept out or so reduced that they do not threaten the trees, the advantage of improved growth alone does not make plastic shelters cost-effective. Except in very small areas, fencing is to be much preferred.

FIRE

Fire danger is a combination of fire hazard and fire risk. The *hazard* is the susceptibility of the vegetation to burn, depending on the kind of plants, the time of year, the wind force and the dryness of the air. The fire *risk* is simply the likelihood of a fire starting, high when many visitors are in the countryside, some of whom are bound to throw away cigarette ends and others who want to have barbecues.

Tree-crown fires are very unusual in Britain. Most bad fires in woodland are soon after conifer afforestation, from seedling to thicket stage, when there is abundant grass or heather fuel and easily ignited resinous tree foliage in easy reach of the flames. In such fires the trees are usually killed and replanting is the only course. Fires in broadleaves may be less disastrous; twigs and stems may be killed but the young trees can be cut to the ground and will send up new shoots, as coppice.

Fire Prevention

The best protection is to prevent a fire starting. Here is a checklist of suggested actions to reduce fire damage in small woodlands.

- In dry periods, ask visitors not to smoke while in or near the woods. If people are working in the woods, prepare areas for them to smoke safely during breaks, cleared down to mineral soil – and enforce the rules.
- If there is a public road alongside the woodland, reduce flammable material there by mowing verges regularly.
- Make fireproof the places where cars may park.
- Brash young conifers to 2m in height to leave no ladder by which a ground fire may climb into the crowns. Along high-risk areas, brash every tree and, on a strip 15m wide along the edge, carry the cut branches back into the stand.
- In some districts there is a tradition of youthful arson, perhaps based on the misconception that early-spring grass fires 'do good'. If this is so in your district, education through schools and community leaders is probably the best approach, showing what is intended in woodland management and, if possible, involving the young people in the planting. Along vulnerable boundaries, reduce potential fuels by making a fire-break of a green vegetation strip (mown grass or an arable crop such as rape, which may encourage wildlife).

Fire Readiness

It is wise to assume that some day, in spite of preventive work, there will be a fire affecting the woodland. There are several ways of preparing; how far to go depends on the level of risk and the hazard.

- Join with neighbours in having a fire plan and agree mutual help. Who has what equipment? Where is the best access?
- Contact the fire-prevention officer of the local fire service so that he may contribute to the fire plan and knows the area. Keep the fire telephone number handy.
- Keep fire tools in good condition and handy. These may range from a shovel and a fire extinguisher in the boot of the car, through a stack of fire-beaters, to making a water tank or pool in the woods, depending on the size of the area and the scale of the danger. (Fire-beaters such as birch brooms at the roadside have a propaganda effect, reminding people of the danger, but they rot in a year, so keep another stock of them under cover for real use.)

Fire-fighting

If fire breaks out:

- Tackle it immediately if it is just starting: small fires quickly grow larger, so do not delay a minute.
- Size up the situation. Where is the fire? What is burning? How intensely? What is the wind direction and strength? These determine where the fire is going to spread. Realistically, can the fire be put out within five or ten minutes with the resources present, or do you need help? If you need help, send for it immediately.
- Weather is vital. A wind of 20mph – a strong breeze – makes fire extremely dangerous. Air humidity less than 30 per cent dries out grass and leaf-litter.
- Fires spread fastest uphill, more slowly on the down-slope.
- The type of vegetation is also vital. A fire in purple moor grass (*Molinia caerulea*) will advance almost as fast as the wind, with burning grass stems leapfrogging the flames ahead as fast as a person can run; such a fire is very life-threatening.
- Fires in leaf-litter are best fought with a

shovel to dig up soil for smothering (beating it may spread it).

- If many people engage in fighting the fire, there must be a chief who takes charge.
- When the fire appears to be out, whether small or large, patrol the area and damp down hot spots with water until it is really finished.
- Take care that a woodland or moorland fire does not smoulder on underground, in peat, leaf-mould or roots. Such fires have been known to break out again after many days or even weeks.
- Take care of people. Fire-fighting is dangerous, imposing extreme stress, with the heat, heavy work and high anxiety.

WINDBLOW

Severe wind may damage woodlands by uprooting trees or by breaking them. Windbreak, when the roots remain in the ground and the stems break about half-height, is financially the more costly but there is almost nothing the forester can do to prevent it. One must simply clear up the mess and sell what can be marketed.

Windblow tends to be endemic in upland conifer forestry. It is worse where rooting depth is restricted or where poor drainage gives roots little grip on the soil, but for many stands the risk of the trees being overturned is almost a function of their growth, given a high degree of exposure; at a height of about 20m or 25m, the leverage of the wind on the tree crown becomes greater than the weight of its root-system, so it overturns. Once an initial breach has been made in the forest canopy, there is a tendency for the damage to spread in a domino effect. Where 'domino effect' windblow develops, it is best to avoid clearing the half-blown trees at the edge of the blow. Windblow generally stops naturally at a change of species or along a

ride where the marginal trees have larger than normal root-systems.

Protection against wind damage involves:

- Regular attention to drainage so that tree roots are not impeded by a high water table;
- Choice of cultivation method prior to planting so that the tree's roots develop all round (not confined to a plough ridge);
- Prompt attention to thinning; in stands exposed to the wind, windblow almost always develops after a thinning which has been delayed beyond the time when it was due silviculturally.

FROST

Damage by frost to natural and major species is usually confined to the seedling stage and to low-lying areas where cold air collects in so-called frost-hollows. At planting, care should be taken to avoid allocating species which are frost-tender in youth to such places; for instance, in a frost-hollow, hornbeam might be planted rather than beech, or Norway spruce in place of Sitka. Good silviculture is the only protection.

MANKIND

Finally, protection is always required from damage by Man. Handsome trees in new housing areas are supposedly valued for providing 'a mature wooded environment' but in fact are often treated with scant regard. For instance, at the building of a famous teacher-training college, two mature beech trees were retained, flanking the entrance. High semi-circular walls were built round each tree (the foundations cutting off half the roots) to give grandeur to the gateway. To enhance further the imposing entrance, the architects then had a metre-high dome of concrete put round each tree, tight up to the bark, the surface decorated with water-worn boulders, and all set in a sea of tarmac.

Inconsiderately but not surprisingly, the trees promptly died.

Trees are made dangerous by having roots cut off by trench-digging for pipes and cables close to their boles, Tree Preservation Orders notwithstanding. Tree roots need air and water and cannot survive for long if the ground is covered with tarmac or concrete. If trees are to be retained, the whole rooting area must be protected during building operations to avoid soil compaction and direct damage, and subsequent landscaping must respect the ground level, neither raising it nor bulldozing it away. Good arboriculture by foresters is not enough; common sense by architects is also required.

FURTHER READING

Bevan, D. (1987), *Forest Insects: A Guide to Insects Feeding on Trees in Britain*, Forestry Commission Handbook 1 HMSO, London

Broxey, J. (1997), *The Potential for Biological Control to Reduce Hylobius abietis Damage*, Research Information Note 273, Forestry Commission, Edinburgh

Coutt, M.P. & Grace, J. (1995), *Wind and Trees*, Cambridge University Press, Cambridge

Dobson, M.C. (1991), *De-icing Salt Damage to Trees and Shrubs,* Forestry Commission Bulletin 101, HMSO

Forestry Commission (1996), *Muntjac Deer – Their Biology, Impact and Management in Britain*, Proceedings of Conference at New Hall, Cambridge, 1993, Forestry Commission and British Deer Society

Forestry Commission (1997), *Phytophthora Disease of Alder*, Forestry Commission and Environment Agency (6pp.)

Gillanders, A.T. (1908), *Forest Entomology*, Blackwood

Gills, J., Brasier, C. and Webber, J. (1994), *Dutch Elm Disease in Britain*, Research Information Note 252, Forestry Commission, Edinburgh

Greig, B.J.W., Gregory, S.C. & Strouts, R.G. (1991), *Honey Fungus*, Forestry Commission Bulletin 100, HMSO, London

Heritage, S. (1996), *Protecting Plants from Drainage by the Large Pine Weevil and Black Pine Beetle*, Research Information Note 268, Forestry Commission, Edinburgh

Ingleby, K. *et al.* (1990), *Identification of Ectomycorrhizas*, Institute of Terrestrial Ecology, Research Publication 5

Innes, J.L. (1990), *Assessment of Tree Condition*, Forestry Commission Field Book 12, HMSO, London

King, C.J. & Fielding, N.J. (1989), *Dendroctonus micans in Britain: Its Biology and Control*, Forestry Commission Bulletin 85, London

Pepper, H.W. (1992), *Forest Fencing*, Forestry Commission Bulletin 102, HMSO, London

Pepper, H., Neil, D. & Hemmings, J. (1996), *Application of the Chemical Repellent Aaprotect to Prevent Winter Browsing*, Research Information Note 289, Forestry Commission, Edinburgh

Phillips, D.H. & Burdekin, D.A. (1982), *Diseases of Forest and Ornamental Trees*, Macmillan, London

Potter, M.J. (1991), *Treeshelters*, Forestry Commission Handbook 7, HMSO, London

Quine, C. *et al.* (1995), *Forests and Wind: Management to Minimise Damage*, Forestry Commission Bulletin 114, HMSO, London

Ratcliffe, P.R. & Mayle, B.A. (1992), *Roe Deer Biology and Management*, Forestry Commission Bulletin 105, HMSO, London

Speight, M.R. & Wainhouse, D. (1989),

Ecology and Management of Forest Insects,
Clarendon Press, Oxford
Strouts, R.G. & Winter T.G. (1994),
Diagnosis of Ill-Health in Trees,
Department of Environment and Forestry
Commission Research for Amenity Trees
No. 2, Forestry Commission, Edinburgh

Multi-Purpose Woodland: Planning and Management

9

The forester should treat time with a disrespect that would appal all but the theologian, who is daily face to face with eternity.

Mark L. Anderson, 1956
BBC Third Programme Radio Broadcast: *Time for Forestry*

The concept of multi-purpose management takes in a wide spectrum of woodlands, all based on the proposition that, by not straining for 100 per cent attainment of any one principal objective, the woodland can deliver a range of valuable secondary products and services at modest extra cost: where timber production is dominant, wildlife can thrive and public access be provided; trees planted as a screen can produce timber; community woodland designed mainly for public access will also be open to many wild creatures. Any of these purposes – and others – may be the dominant objectives and in most circumstances others can be accommodated. Sometimes, although it is seldom, the single-purpose woodland may be appropriate. More often the duty and skill of the forester is to see what uses are compatible and how to achieve the best mix.

The principal duty and interest of the true forester is not to exploit the forest but to ensure its perpetuation. He must so control those who exploit it that the harvest is fully balanced by growth and the harvest damage is repaired. In some countries well endowed with natural forests, notably in Continental Europe, trained foresters have been able to put in place conservative management systems capable of ensuring that these semi-natural resources can yield their harvests of wood and other benefits in perpetuity. In contrast, all too often in the last half-century in developing countries around the world and in spite of good management controls put in place by skilled foresters, the temptation of quick profit has proved too much for local politicians and ruthless lumber companies, so the controls have been discarded and the forest destroyed. In Britain the natural forests were destroyed long ago by the exploiter and by ill-conceived land-use change; the main task of the forester now is to recreate productive woodlands lost and to rescue to sustainability the relics we have inherited.

These are long-term affairs and it is all too easy as time passes for intentions to become obscured and reasons for tree-growing decisions to be forgotten. Woodlands need stability of purpose. Both in national policy-making by politicians and in silvicultural policy locally, forestry benefits from continuity of practice and the avoidance of violent change.

A major help towards consistency is

achieved by writing a management plan and then working to it. It may be brief or elaborate, depending on the size of the woodland and the complexity of the problems and opportunities it presents, but the action of writing a plan often clarifies the issues and improves the manager's decisions. The plan, once made, becomes a commitment, at least as a self-discipline, and it is now a requirement for the payment of some Forestry Commission grants that a plan has been prepared and approved.

The records of work which develop from the implementation of a plan are of immense value in woodland management, showing the result of previous decisions and providing the means of avoiding the repetition of mistakes. Where did the seed come from for this stand? Were these oaks planted pure or with nurses? What thinnings were done? What timber volume was cut last rotation? Why on earth did they plant this mixture? The answers to such questions should allow you (or your successor) to do better next time.

The operational plan should prescribe the programme of work for a period of five years. The descriptive part will be largely unchanged from one five-year period to the next, so most has to be written only once. The whole management plan should be succinct and kept to essentials as a working document. Whether two pages or something more elaborate for a large woodland, the plan essentially consists of four parts:

1. Description
2. Objective
3. Prescription
4. Record of action.

DESCRIPTION

At its simplest this will name the wood and provide a marked map to show its boundaries and location. However, since the facts about the environment are the basis of so many decisions on what trees can be grown and how best to tend them, it is useful also to describe briefly the soil (its structure, fertility, drainage, etc.), the elevation above sea level, the rainfall, the steepness and aspect of the slopes and the exposure of the site.

This section must include a description of the present stands of trees: the species, their proportions, age and size or development stage, and their condition (or the land may be bare of trees), and also the ground vegetation and bushes. The actions of writing a note about the plants – which are indicators of fertility – and the proportions and present condition of the various tree species will cause the writer to look critically at what is really there, as distinct from a casual glance or – what is worse – an assumption.

Table 9.1 contains a check-list of possible items for description, as the basis for the plan; some may be one-line answers and some of the headings may be unnecessary, but the writer should consider if a note on any might be helpful in deciding among options now or for someone 50 years hence in understanding why you came to a particular decision.

Table 9.1

DESCRIPTION OF WOODLAND FOR OPERATIONAL PLANNING

- Name, location; map or sketch with boundaries
- Land area, with divisions if they clarify management prescriptions. Boundaries stock-proof or open to grazing?
- History and past management. Is this an ancient woodland? Any archaeological remains?
- Rainfall and elevation above sea level
- Soils, drainage, slopes and aspects
- Trees and shrubs: species, proportions of each, ages, stocking, health and condition
- Ground vegetation: dominant species and

Caledonian pine forest conserved: seedlings, poles and mature trees in the Black Wood of Rannoch

Caledonian pine forest in dire trouble: no regeneration for many decades owing to sheep grazing, Glen Falloch, Perthshire

Silver birch

A timber stand of Douglas fir

LEFT: Group regeneration in conifer woodland; *RIGHT*: A tree walled in and given a concrete necklace: the architect was paid but the tree died

Well-managed peri-urban woodland: pole-size regeneration of oak and other broadleaves

A famous designed landscape: Scott's view on the River Tweed

Avenue planting of common lime flanked by younger beech, Ballathy, Perthshire

Homegrown: roof timbers grown and sawn on the farm

A mobile sawmill in action: bandmill towed on site by Land Rover
(Forester Company, Standrange Ltd)

Charcoal-burning: modern steel kilns

Short-rotation coppice: three-year poplar with pipes laid out for future
application of liquid slurry (© Border Biofuels Ltd)

Coppice-with-standards: hazel coppice ready for cutting under an oak standard, the shape typical of this system

any unusual ones

- Birds and animals: presence of rare or unusual species? Rabbits, deer, grey squirrels?
- Game-shooting
- Statutory designations (Site of Special Scientific Interest, National Nature Reserve, Special Area for Conservation, National Park, etc.)
- Special landscape considerations
- Existing public access, whether right-of-way, privileged paths or *de facto* access
- Special constraints on management: fire risk, air pollution?

OBJECTIVES

What do you intend the wood to provide? Shelter for farm stock; growing quality sawlogs; firewood for home; pheasants for shooting; habitats for wildlife; space for people to walk; screen for the council land-fill tip? It is likely there will be more than one objective and, if so, their order of importance should be thought out and written down. Once again, the action of putting these intentions on paper may focus thinking on what are feasible objectives and what are unrealistic combinations. The rank order of importance given to multiple intentions may influence the choice of species to be planted and the way they are tended.

A useful section of the plan, especially in retrospect, should follow the statement of objectives: that is, your reasoning in combining the facts of the description with the intentions for the wood. This will result in the rejection of some species and some systems of silviculture and a judgement in favour of others. You draw on the evidence of successes and failures, including those in neighbouring woods, to declare how you intend to achieve the aims.

Finally, it is of great value if you describe succinctly what is the silvicultural objective,

i.e. your concept of what the woodland should look like in the long-term if all goes well, both in species composition and structure. For instance, you may be planting a wide mixture of species for short-term cropping for income and as nurses, but your intention and vision is that the wood should ultimately comprise, say, a high forest sessile oak (80 per cent of cover), birch (10 per cent) and mixed broadleaves including bird cherry, aspen, gean and rowan (10 per cent), not all of which may be present at the original planting or when you write the plan. Your successors – and indeed contemporaries – should know that.

PRESCRIPTION

This section comprises instructions, or orders for action, and must be written directly. It will be divided into annual schedules of work: cut this; fence that; plant these species in these groups or areas; weed that. Each operation must be quantified so that the amounts of materials, plants, labour, etc. can be immediately worked out. There should be no need for description or justification; all those have been given before.

As a matter of convenience, foresters divide large woodlands into compartments, to reflect lasting differences of site and of silviculture and so that prescriptions may be precisely located. As far as possible, compartment boundaries should follow natural and permanent features; straight-line divisions through the middle of a wood are seldom satisfactory. Compartment divisions should be included in the description section and shown clearly on the maps.

RECORD OF ACTION

It is easy to have good intentions – the road to Hell is paved with them. For the woodland plan, however, honesty requires a strict statement of what has been achieved. For convenience this section should comprise a form for each year; the left half should set down in brief what is prescribed, while the right half has blank space to record the work actually done, to be completed each year. There should also be space for a brief explanation of any discrepancy between the prescription and the achievement.

The need for written plans and written records of the actual achievements against the prescriptions is explained by the long life of trees and the long production period of woodland. It is remarkable how quickly and how completely incidents are forgotten which, although quite trivial in themselves, have a lasting effect on a woodland, such as one night's frost or half a dozen sheep forcing an entry to the young plantation. Committing the incidents to writing is the only way in which someone in 20 years' time may know why the stand is so poor – or in other circumstances why it is so good – and may be prevented from drawing a false conclusion, often benefiting from the knowledge.

Finally, the record section should include a reminder, to be acted upon at the end of the fourth year, that the plan should be revised in good time for the beginning of the next five-year period.

How the management is organised depends greatly on the silvicultural system adopted in the woodland. In a clear-felling system the various operations will be done as a sequence in any block or compartment: felling in one year; next year preparations for replanting; planting; next year beating-up and weeding; years later cleaning; later still thinning. In contrast, in a group-selection woodland all the operations will be done in the block of the year. Within that block there will be groups felled and other groups cleaned and others thinned – the full sequence of work is concentrated in one block of the wood, although in separate groups, since all ages from regeneration to maturity are found together. Each block will be visited on a cycle of about five years.

Woodland management thrives on continuity of purpose and requires patience. In economics the term 'short-run' has a special meaning: it is the period in which an industry's production is restricted to the present set of factories or production units; the short-run ends when new factories can be built and brought on-stream, perhaps in four or five years' time in many kinds of business. In forestry, in contrast, to create and bring on-stream a new oak forest will take at least 120 years and even a new Scots pine forest may take 70 years. That is the length of forestry's 'short-run'.

It comes as a jolt to others to find a forester in charge of coppice woodland run on a ten-year rotation looking forward quite naturally to the normal harvesting of the same area ten years hence, and 20 years and 30 years. What is an obvious and routine projection to the forester appears to stagger his friends as an incredibly long forward view – predicting 20 years ahead!

Twenty years? The time-frame for forestry planning is longer than that for most important stands. If we should fail nationally to sustain the mainly production forests of conifers there will be a gap in wood supply to mills in 40 or 45 years' time, yet the assurance of future supplies is a prerequisite for investment in pulping and sawmill machinery and hence for employment in the industry. We have a duty to sustain the woodlands of Scots pine, oak and other broadleaves, so that all the benefits of these multiple-use resources – timber, wildlife, environmental protection – may be available for people 100 or 150 years hence and more. If you have any doubt about that responsibility, or if the return seems too remote to be real, stand by a big tree, or better still in a mature wood, and ponder the fact that we have inherited these from someone 150, 200 or even 300 years ago. Take the opportunity to plant trees and sustain woods now and 100 or 200 years hence someone may think well of you, even if they will not know your name; there are not many things we may do in this world to achieve that.

The urge to sustain the yield of the forest comes not only from its owner and those who benefit from it directly, but from the buyers of wood and the managers of mills who want assurance that the raw materials they need will be available regularly in the future.

There are two ways of conceiving the management of forest resources, fundamentally opposed. On the one hand there is the plantation concept, that the land is the basic fixed capital on which crops are raised, each separately, so that when each plantation is mature, it is harvested and the manager considers what should be done next on that land. Directly opposed to this, it is properly held that the forest's fixed capital comprises the land and the growing stock on it (trees, bushes, plants, animals, etc.) which can be continually and perpetually cropped, regularly yielding its harvest, ever changing in detail as stands grow, are cut and regenerate, but never changing in aggregate. The former philosophy may produce excellent plantation crops but has no commitment to sustention. The latter is the attitude of the forester.

Since the days of man-the-hunter-gatherer, it has been recognised that woodland can yield much more than timber – fruits, wild game, domestic grazing, soil-erosion control, recreation and so on. Often, however, the harvest of wood is the only product which sells in the market to provide revenue. The management of woodland which produces multiple products, some priced, some unpriced, has long posed – and continues to pose – sharp difficulties, because the roundwood, the unprocessed wood as it leaves the forest gate, is not so valuable that it can easily carry the costs of producing the other goods and services which most people expect to enjoy free.

The problems of joint product management are not confined to forestry, although the forester faces them more harshly than most. Increasing investment brings the temptation to concentrate more exclusively on the products which are marketed and which pay best, and to eliminate products which are unmarketable and which are, in the accountant's view at least, leaks through which the result of investment is lost and wasted.

In Britain the national policy favours multi-purpose woodland, both in the Forest Enterprise and in the private sector. It is seen by most people as 'a good thing' that woodland should produce wood for sale *and* be managed in such a way that habitats for wildlife are provided, along with a range of other unpriced services. In a normal manufacturing business, the prices of the various products determine how much of each will be produced, but in the forest business, with some products priced and some unpriced, this cannot be the decision mechanism. Where several woodland products are saleable, some combination of joint products yields a total income greater than any single crop. But where there is a mixture of marketed and non-market products, the cost of producing the unpriced services must be carried ultimately by the timber which is, usually, the only saleable product; alternatively, there must be a subsidy (private or government) to bridge the money gap. Foresters can design and provide multi-purpose woods, but these require a sacrifice, at least an apparent one in the first rotation, compared to the strictly functional and least-cost tree-farm.

Commitment to multi-purpose woodland may yield a lower financial return on the investment than might be attained by concentrating solely upon the product with the highest market price. There are offsetting benefits, however, in flexibility and perhaps in ecological robustness, so that the penalty may be illusory. A mixed woodland in which the species are well matched to the local soil and site factors is likely to suit wildlife, including game birds, and to look well in the landscape – given care with its size and shape. Furthermore, it is generally easier with mixed species to serve a range of markets, including produce from the early years of the rotation, which should be a feature of small woodland management. It is impossible to forecast with accuracy what demand for particular timbers will be 50 or 100 years hence. It is clear, however, that demand for wood throughout the world is increasing and the supply, especially of high-quality woods, is ever-diminishing.

As a result of pressure to reduce costs by cutting labour inputs, the management of large forests has been forced to become extensive and to concentrate on the bulk production and marketing of a uniform product. This coniferous wood, sold as the raw material for paper-making or particle-board, may be transported 200 miles or more from forest to mill, generally in 38-tonne loads. To the grower, the profit per tonne is small and the trade is attractive only by virtue of the large amounts involved. It is not a management strategy that is advantageous to the small woodland grower who should not seek to emulate it.

Instead, small woodland is best designed and managed to produce high-quality goods and services and, if possible, to serve a diversity of local markets, minimising transport costs (thus leaving more for the grower) and encouraging the establishment of local businesses to process and add value to the raw material. In this there is good potential for the creation of co-operative marketing among neighbours and even for the sharing of machinery. Through such co-operation a market may be created which would not otherwise arise, since no single small grower may be able to give an assurance of annual continuity of supply. Co-operation in marketing and sharing machinery are less difficult in forestry than in farming; wood harvesting is far less seasonal and weather-dependent than cereals.

Above all, the small woodland forester should aim at high quality, not mere quantity production.

The first priority for woodland management should be to ensure that existing woods are brought into good order; even if they are in poor condition at present, they represent an accumulation of many years' woodland capital, which should be safeguarded. Woods in a derelict condition command no respect from people and tend to attract anti-social behaviour, including the dumping of rubbish. In contrast, woods in good order, where improvements are seen to be in hand, tend to gain people's respect and support.

The sustainable management of woodlands presents a severe challenge, arising from the fact that the annual growth of wood cannot be identified as an entity separate from the forest capital (as a crop of lambs is distinguished from a ewe flock), but is laid down – like a coat of paint – over the outside of every tree in the forest. If the woodland is being managed on a sustained-yield basis only the equivalent of this amount, which is difficult to measure, can be removed; to cut less would be safe but unthrifty and would begin to make the woodland overmature, and to cut more would eat into the capital. Since the actual growth cannot be harvested separately, the forester must estimate it and aim to harvest whole trees up to that amount and no more. Although this issue may arise directly only in larger forests where sustention is possible on an annual basis, the same principle applies in smaller woods where yield can be sustained periodically. The manager may control the harvesting simply on an area basis, for instance in coppice, where an equal area is cut each year; or he may control it by marketing equal volumes of timber, which is more effective for customers but more difficult to achieve. In practice there must be frequent departure from the ideal harvesting volume because of wind damage or fluctuating market demand, but the principle and safeguards remain important if the woodland is not to be open to irresponsible or unscrupulous capital-stripping.

Sustention – that is, management which sustains all inputs and outputs – means that the woodland has the capacity to provide its goods and benefits indefinitely, to the twenty-third century, the twenty-fourth, the twenty-fifth . . . there need be no end.

At the start of this chapter, a distinction was made between the work of bringing into conservative management woodland-in-being and that of recreating woodland on bare land. The difference lies not only in the nature of the work to be done, but in the forester's attitude to time.

For the woodland which is in-being and already under sustained management, time is of no account. Regeneration and harvest are contemporaneous (albeit in separate areas); there is no waiting, because this is a steady-state system in which all costs and all benefits are incurred and received continuously.

In situations where the woodlands must be created by planting or where derelict woods must be patiently restored to good order, time assumes great importance because, nationally or as individuals, we must wait for trees to grow, from seedlings to maturity. To accountants, however, time is money, and bankers and Treasury try to insist that interest is paid on the planting investment, usually at a rate higher than the natural growth yield of trees.

The counter-argument is that years ago society pocketed the proceeds from its exploitation and destruction of the natural woods, and, if there are to be woodlands in future, society should now pay back sufficient for their re-creation. For accountants and economists it is a great difficulty that some of the major benefits provided by woodland are not paid for in money in any market and therefore cannot be properly valued; there is no common currency which measures wood, woodcocks and the sanity people gain from quiet recreation. Rather than try to evaluate what is invaluable and get it wrong, perhaps it may be best to leave unspecified the value of benefits which are truly imponderable.

What is clear is that there is no 'quick fix' for creating woodlands or national forest resources. In 1956 Mark Anderson suggested the following:

It is not so much time that the forester requires for successful forestry as patience, tenacity and skill on his own part, and patience, encouragement and intelligent support from others. To secure these last, the forester must disclose his problems. He should not deceive himself, nor the public, by minimising them. When the public understand the position, they will not hustle him – they will allow him plenty of time for afforestation and all of future time for forestry.

REFERENCE

Anderson, M.L. (1956), *Time for Forestry*, BBC Third Programme Radio Broadcast.

Woods for Wildlife

10

The natural forest is the most diverse of habitats . . . the climax forest where the trees are of mixed species and mixed age, with areas of extreme density grading into open woodland and natural clearings.

Frank Fraser Darling and J. Morton Boyd
The Highlands and Islands, 1969

This quotation pictures one ideal for a wildlife wood, but in Britain we have few areas of natural forest left and must try to mimic various types to provide habitats for some wild creatures which are hard pressed or which we choose to favour; simply waiting for nature's climax is scarcely an option. Although most woodlands are managed for multiple purposes, some – or at least parts of them – may be designed principally as habitats for wildlife. The idea may be to help a wide spectrum of plants and animals as a refuge for any wild creatures in an otherwise intensively managed countryside, or the aim may be to provide good conditions for a specialist group, such as butterflies, or for a single species, such as the pheasant. Forests grow slowly, therefore it is wise to think closely about the purpose so that management may be consistent. If woodland is to serve wildlife well it must be designed and managed; a regime which will suit some plants and some creatures will not suit others.

It is not possible here to give detailed prescriptions to satisfy the special needs of every group and every type of woodland – direction to further reading can do that more satisfactorily – but general principles can be indicated.

Under the closed canopy of most trees such as oak or pine, only a modest cover of bushes and ground vegetation can grow, and perhaps none at all under dense spruce or beech, because in northern latitudes the amount and quality of light penetrating full crowns of those trees is inadequate for active photosynthesis at ground level. Most insects and all grazing animals depend directly on plants to capture solar energy, so the basic requirement in specialist woods for wildlife is that there should be actively growing plants other than tall trees.

The structure of the woodland is the key to sustaining actively growing shrubs and ground plants. A well-stocked even-aged wood, while it is young, can support a huge amount of wildlife for a limited period, but it may then grow past that phase into relative wildlife poverty. For instance, the large pure conifer plantings of the 1960s and 1970s in south Scotland held extraordinarily large populations of roe deer from about year five to year 20 after afforestation; the deer bred fast and grew to international bronze and silver trophy standard, because of the high-quality feeding and the shelter available for those years. Then the spruce canopies closed, the conditions changed and the roe populations settled at lower densities. Such an ecological boom and bust economy is not acceptable for a wood dedicated especially to wildlife conservation; the habitats need to be sustained.

A range of habitats satisfactory for many creatures can be fully sustained in woodland managed on a group structure, either group clear-felling or group-selection, or in regularly worked coppice. Many birds and mammals quickly adjust their distribution and home ranges to the habitats which are continually created. This structural arrangement is satisfactory for creatures which are mobile, provided the regeneration areas are not too far separated.

For animals which are not great travellers, however, and for many plants, a system which provides suitable habitats only on a rotation of several years may be inadequate. For instance, the heath fritillary butterfly has been shown to be unable to colonise ideal habitats more than 600m from existing centres. For these creatures, habitats must be provided continuously by annual or short cyclic management, or created within easy reach of colonists.

Many animals, including birds, if they are to thrive, require access to several sharply contrasting conditions, preferably in close proximity: good feeding grounds, cover to hide from enemies, nesting or den sites with sun warmth (especially for cold-blooded animals and insects), freedom from disturbance and shelter from wind. Many of these conditions are well met at the transition zone between woodland and open country, which ecologists call an ecotone; within five or ten metres a creature can find open farmland, trees to shelter from the wind, cover to hide from enemies, a choice of sunshine or shade, a variety of vegetation from woodland plants to sun lovers and the lower organisms which live with each as possible food. Many animals and birds find the woodland edge the ideal place to live; this fact is especially helpful for the management of woods for multiple purposes, since much of the interior may be devoted to timber production.

The design of the woodland edges is essential to success in helping wildlife. Making the edge scalloped instead of straight increases the length and shelter effect, and hence the wildlife-carrying capacity. For many creatures useful edges can also be created inside the wood by forming wide internal rides and glades.

For wildlife conservation, rides in the interior of woods have to be sufficiently wide to allow summer sunshine to reach the ground for several hours each day. This is better provided by east-west rides than by north-south-oriented ones, and they must be about 35m wide for a 25m-tall tree stand in southern England (rather more in the north) to provide a much richer fauna. Internal rides for wildlife should not extend to the boundary of the wood, otherwise they become wind funnels, but should be blocked with trees at the ends. In a large multi-purpose woodland a wide ride can incorporate a road and the additional space will be useful for occasional stacking of logs, so, to some extent, it will be multi-purpose space. It is a help to form south-facing bays on the north side of existing rides; bays 30m long by 6m deep are good, giving extra sun and shelter.

When it is intended to widen rides in existing woodland or to cut bays to create sheltered areas, it is advisable to make a trial cutting one year and to monitor the results on the vegetation before doing the full-scale work. The effect of cutting the trees and opening up the ground vegetation may not be what is intended; there may be a large store of weed seeds from a previous phase of agricultural land use, which will germinate and swamp the woodland-edge plants which are desired for butterfly habitat. Some additional management of the vegetation may be called for, perhaps mowing to prevent these plants reseeding, in order to encourage the woodland ones.

Thought should be given to the size and stability of woodland before cutting a wide ride through it. If the stand is young, has been well thinned and is on deep-rooting soil, there may be no ill-effect. In older stands, underthinned or on shallow or wet soil, the cutting may destabilise the remaining trees and so impair the value of the woodland.

The vegetation in rides has to be managed to sustain its quality as wildlife habitat; the herbaceous vegetation in the central two or three metres may best be cut annually to provide access; adjacent strips on either side, often with small shrubs, may be cut on a three- to five-year rotation, the cutting years being staggered on the two sides of the ride and also in adjacent lengths. Ideally the cuttings should be removed; brushwood can be taken into the wood and put into piles as additional wildlife habitat.

In woodland managed as high forest for multiple purposes, useful wildlife benefit can be achieved by linear coppicing along the rides, coupled with the creation of 'box-junctions' at the ride and road intersections to allow more sunlight to penetrate. This can create useful habitat along which insects in particular may move to colonise the box-junctions.

While management of wide rides and external edges is the preferred strategy for

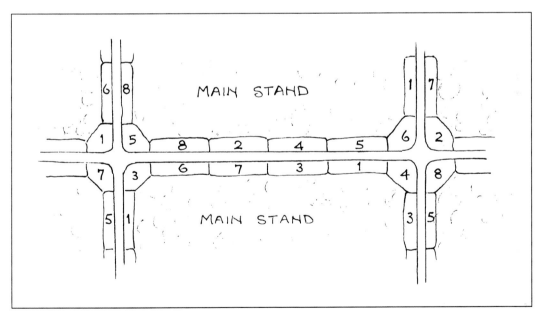

BOX-JUNCTION MANAGEMENT, for butterflies etc.
Strips of coppice 10m wide edge the roads, cut – in this case – on an eight-year cycle, which allows plants tolerant of periodic shade (e.g. viola) to thrive. Sections of the strip and the broader box junctions are cut in sequence each year so that insects easily migrate to find ideal feeding and breeding conditions. (After J.N. Greatorex-Davies in Forestry Commission Occasional Paper 28)

wildlife conservation in conjunction with high forest, similar results can be achieved with coppice or coppice-with-standards management. Part of the value of coppice lies in the fact that the tree canopy becomes sufficiently dense within its short rotation to shade out and kill the typical open-site plants, the thistles, stinging nettle and rampant grasses such as tufted hair grass (*Deschampsia cespitosa*). The shade-tolerant species such as viola and primrose survive the closed-canopy stage and are poised to flourish, spread and set seed when the coppice is cut and when, for two or three years, there is abundant light. These conditions are ideal for many of our less common butterflies and for a host of other insect species upon which birds may feed. One condition for the practice of coppice management for wildlife is that the cutting areas or coupes should be of modest size and reasonably close in successive years, say within 300m, in order to allow relatively immobile insects to colonise each area when it is cut and as the previous one becomes less attractive. Individual coppice coupes may suitably be about 0.5ha, with a rotation of eight to 12 years. Coppice coupes in a coppice-with-standards system may be slightly larger but should not exceed 1ha. A structure especially suitable for a wildlife wood is a mosaic of coppice coupes and groups of high-forest trees, which may be easier to manage than coppice-with-standards; it is a modification of group clear-felling and the coppice coupes should not be less than 0.25ha.

The management of woodland for wildlife almost always increases the amount and quality of browse – of shrubby growth – and the inevitable consequence is an increase in the numbers and breeding performance of woodland deer, principally roe deer in most parts of Britain, although maybe also fallow, muntjac, sika and even red in some parts of the country. (Sika deer, introduced from Japan, appear to be actively replacing roe as the specialist woodland deer where the two species come into contact, as in the Scottish Borders.) There may be such an increase in deer numbers that their highly selective browsing and grazing radically changes the vegetation, including preventing tree regeneration. In the absence of large carnivores – we have none in Britain – deer may take such advantage of the improved conditions that they defeat the main purpose of the management, say for butterfly populations or pheasants. It may be necessary, therefore, to control the deer by shooting; concentration of cutting in coupes for coppice management makes this much easier than the linear systems, since the vulnerable vegetation of each coupe can be covered from one or two high seats. If the deer population exceeds about eight small deer or four red or sika per 100ha, the animals will almost certainly restrict tree regeneration and the effects of their selective feeding on other palatable plants will be evident. Deer control will then be essential in the interests of conservation. Being naturally secretive, deer in woodland are exceedingly difficult to count, and even experienced ecologists habitually underestimate their numbers; there are almost certainly more deer living in your woodland than you think.

Areas specifically for deer require careful planning. Open glades between 0.2ha and 0.8ha are probably best, preferably on a sunny aspect. Glades surrounded by trees offer shelter from wind and for that reason are rather better than long rides for holding deer. They should be left unplanted with trees and the ground vegetation may be managed to improve the grazing, perhaps by fertilising, liming and reseeding with native grasses and clover, or perhaps with field crops such as rape for red deer. Roe deer especially are browsers rather than grazing animals, however, and they are attracted by palatable browse plants; willows are special favourites. All these glades require management to maintain them in an attractive state, periodic mowing of the grass (removing the cut material from the site) and coppicing the willows. Viewing hides may be built at some glades if the wildlife wood is also to serve education. Glades are essential for deer management, for both counting and culling.

The treatment of the woodland edges is critical for the success of wildlife woods. Shrubs and small trees should clothe the edges so that they are sheltered and warm, not draughty and cold. A sudden transition from the bare boles of tall forest trees to open farmland is inhospitable for wildlife, whether songbirds or pheasants. The shade cast by tall trees growing right up to the boundary weakens or precludes the growth of shrubs. The most desirable structure is one with a shrubby edge or hedge, backed by small trees, and these by the main-stand trees.

The edge may be of willow, hawthorn, dogwood, holly, rose, guelder rose, elder, hazel, wild service, sloe and similar shrubs and a hedge, if that is preferred, of beech, hornbeam and holly. Choosing a hedge means there must be periodic cutting so that it does not get out of hand. Species suitable for the retro-margin zone include hazel, field maple, wild service tree, rowan, Swedish whitebeam, aspen, gean, bird cherry, crab apple, willows, holly, yew and sea buckthorn, the choice depending on the locality.

The choice of the main tree species depends on the local site conditions but species which do not cast heavy shade are appropriate, including oak, birch, ash, pine, larch and cherry. Beech is less desirable because its shade greatly reduces the vigour of shrubs and ground vegetation.

This type of edge is quite suitable for

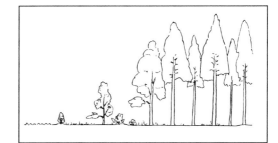

EDGE STRUCTURE

The marginal zone, 1 to 1.5 mature tree heights in width, contains spaced native broadleaved trees and shrubs, a boundary hedge for warmth and open grass. This is robust, helps to reduce windblow risk in the main crop and gives a wide range of feeding, breeding and roosting habitats for wildlife.

pheasants, the principal management objective for many small woods in the lowlands of Britain. Special features which game-shoot managers would want to add would be roosting trees and some evergreen conifers to improve the wind shelter in winter. The roost trees could be larch and spruce, both of which have abundant horizontal branching at a height safe from foxes; they should form at least 5 per cent of the main stand at maturity. The evergreens might be Norway spruce, Lawson cypress and silver firs in the main stand, and cotoneaster in the shrubs.

Food and habitat quality limit the number of pheasants a woodland can hold, whether wild or released birds; even released birds will not remain in an unsuitable wood. The woodland edge is the critical habitat for holding pheasants, especially to give good cover for hiding and ready access to feeding areas. Any shrub-planting to improve the habitat for pheasants should be concentrated along the woodland edge before any effort is spent underplanting in the interior. For pheasant-shooting it is the condition of external edges that is important; internal rides are secondary. The greater the length of good-quality edge available, the greater the number of pheasants that can be held, so small woods, say 3–5ha, and long, narrow woods which have long external boundaries to their unit area are desirable for pheasants – but perhaps less so for other creatures.

Improvement of woodland for pheasant-shooting is quite compatible with improvement for butterflies and other wildlife. Conflict of interests may very well arise, however, if general wildlife woods are used for intensive pheasant-rearing, since the area around the rearing pens is likely to become infested with weeds foreign to ancient semi-natural woodland and to be affected by trampling and nutrient-enrichment. Another problem may arise from the practice of spreading straw on rides used as feeding areas, leading to the smothering of plants, nutrient-enrichment and the introduction of weeds. Although these are conflicts in ancient semi-natural woods, they may be acceptable in new plantings and may be reduced or avoided by careful management and feeding from hoppers.

For the encouragement of wild pheasants (and of many other birds), the juxtaposition of the woodland with rich feeding is crucial. After hatching, pheasant chicks are led to insect-rich areas for feeding, either on an adjacent cereal field or on newly cut coppice. The appropriate management of internal rides by rotational cutting can provide good habitat for insects, which automatically favours insectivorous birds and bats, provided that roosting and hibernating sites are also available.

The provision of wide, sunny rides and coppice management are likely to maintain good populations of small mammals, especially voles and shrews, which in turn maintain weasels, foxes, stoats, kestrels and owls.

Mature and over-mature trees are important as hosts for bark-living lichens. Oak and ash are good in this regard, as, in the north at least, is the sycamore.

Deadwood is an important element of the forest ecosystem, supporting many specialist wildlife species, some of which have become rare, as few woods are now allowed to accumulate the amount of deadwood typical of a natural forest, which may be as much as 100m³ per ha. In high forest managed for timber the equivalent amount may be less than 5m³ per ha. Piles of brushwood up to 1m high provide a rich habitat for insects and, in consequence, for insectivorous birds such as the wren.

Oak and elm (the latter the victim of Dutch elm disease) are the most important providers of deadwood, being large, decaying slowly and supporting a greater diversity of organisms than fast-growing and fast-decaying non-native trees. Birch, which reaches maturity early and rots quickly, is among the less useful deadwood suppliers but is obviously important in upland forests where it is a common component. It does host several fungi.

Dead branches and rotten heartwood of living trees provide special breeding sites for rare wasps, beetles and flies. Larger holes provide nest holes for many birds, from redstarts to owls. Rot holes with water are breeding places for certain hoverflies, some known to breed only in these places. The fallen deadwood supports a wide range of insects and fungi, from large longicorn and rhinoceros beetles to mites and springtails, and many wasps and bumble-bees hibernate in holes within it. Not least important, the fallen debris on the forest floor is the life-material of the mites and worms which, although unspectacular, are the food of many higher organisms in the ecosystem. In wildlife woodland where deadwood is retained, it is important that standing trees do not constitute a danger to people.

Dead trees may also provide roosting sites for bats, of which Britain has 15 species, all protected, all at home in woodland and several quite dependent on it. All British bats are insectivorous and thrive on hunting at woodland edges and along rides. Barbestelle and Bechstein's bats are specialist woodland species. Tree holes may be used as roosts, one of the wildlife benefits of leaving some over-mature broadleaved trees to remain as hollow, dead and dying. Where winter or summer roosts for bats are few or absent, artificial ones can be provided. Winter roosting boxes must be thick enough to protect the animals from the frosts; they should have a central chamber about 100mm square and 300mm high, with walls, roof and base not less than 100mm thick if made of timber – which is not an easy construction to achieve. Summer roosting boxes are lightweight, with an open bottom, conveniently made of wood 25mm thick. Roost boxes may be set high up, three to a tree, in groups of five to 12 trees. As well as the edges of broadleaved woodland, waterside is an appropriate site for encouraging bats. Freedom from disturbance is crucial for them.

Woods for wildlife should give special attention to the restoration of good habitat along streamsides, in both the lowlands and the uplands. In the more intensively farmed parts of the country, in the lowlands and river valleys, most of the land apart from the streamsides has been radically changed, so these are the last refuge of locally native species of plants and invertebrates.

In upland farms where streams are usually unfenced, grazing over many decades has destroyed the woodland and tall herb vegetation which naturally bordered the streams. The loss of the litter and insects which formerly fell into these waters has greatly reduced their nutrient levels; the absence of dappled shade has changed the streams' temperature regime and the erosion of the unprotected banks has filled the deep pools to create braided channels. The result has been a drastic reduction in the productivity of fish stocks in the rivers of which these upland streams are the breeding grounds, and this has been greatly exacerbated by the deposit of acid in rain and mist derived from car exhausts and industrial pollution of the air. Former afforestation practices made the damage worse, with plough erosion and the planting of heavy-shade-casting trees to the very banks of upland streams, but these have now been stopped.

The aim should be to have an effective buffer strip of thriving vegetation on both sides of both small and larger streams. For several years the prescription for afforestation has been to keep planting at least 5m back from the banks of small streams; the real objective of that was to limit the erosion of soil into the streams by having no plough furrow closer than 5m. It is clear, however, that vigorous forest trees growing only 5m either side of a small stream will over-arch it and blot out the sun long before they reach even 20m in height. If these streams are to be biologically productive – and the upland streams are vital to fisheries as the breeding grounds of salmon and trout – they must have sunlight and herbaceous vegetation on their banks to support insect food. The planting along the banks should be designed to let in the sun; on the south side especially, tall timber trees should be kept back 20–25m, and everywhere the treatment might usefully follow that already suggested for woodland edges facing farmland: some low shrubs near the stream, backed by a zone of small-stature trees, these in turn backed by the timber trees.

Approximately half the water surface of the streams should be open to sunlight and the other half lightly shaded by the foliage of mainly native trees and shrubs. Domestic stock should not have access to stream banks because of the erosion damage they inevitably cause; the ideal is to have the banks clothed in tall herbs which should contribute seeds, plant litter and the associated invertebrates – insect larvae etc. – which are essential contributions to the stream ecology and hence to fish production.

In wildlife woods and in timber-production forests, there should be a total ban on tractors driving through streams and on timber being dragged through them. In harvesting operations stream banks must be protected and streams bridged.

GENERAL DESIGN OF WILDLIFE WOODS

In general, the designer of a wood for wildlife should begin by reviewing the qualities of the whole site and selecting the most suitable areas for management towards the needs of different groups, reinforcing the natural strength of each part. The successful strategy is usually to practise general conservation management for the encouragement of a wide range of plants and animals and, within that policy, to major on one group of organisms which are particularly at risk or for which the wood is particularly suited. Close consideration should be given to the relation of this wood to others and to other wildlife refuges, as well as to the maintenance or creation of linear habitats such as hedgerows or woodland strips along which animals may migrate.

One of the greatest attributes of woodland habitat for animals, including birds, is the freedom from disturbance which they can provide. In this respect, the value of forests which many conservationists regard with scorn should not be overlooked. The interior of coniferous forest blocks in exposed sites and designated as 'non-thin' stands may be not only undisturbed but virtually unvisited for periods of 25 or 30 years; since these invariably have patches of failed trees and small windblown gaps, they provide sanctuaries which are almost unparalleled in the modern countryside. That is not all: recent ecological research has given an unexpected twist to this fact. It was long assumed that plantations of exotic conifers such as spruce and Corsican pine were particularly poor in biodiversity; now it has been found that the interiors of such woods (but not the disturbed edges) contain very large and diverse populations of insects and spiders, numbers higher than those found on broadleaves in either temperate or tropical forest. So much for the old sterile image of dark conifers!

All will not go according to plan in wildlife woods. The 'wrong' plants will flourish; natural predators will not respect the manager's wishes. One local conservation group put up 75 nesting boxes designed for pied flycatchers and were successful when all 75 were occupied by nesting flycatchers; equally successful were the pine martens which subsequently raided all 75 nests. Such developments must be accepted. The manager should have a policy on selective interference, particularly if habitats are being engineered selectively to favour one group of creatures. What interference is acceptable and what is unacceptable? Should grey squirrels be destroyed in the interest of protecting the native red squirrel and timber trees? And, if so, what methods are acceptable and what are not? If grey squirrels, what else? Crows, magpies, feral mink or even a 'desirable' creature which becomes excessively common?

These issues, superficially straightforward, often raise difficult questions, particularly in view of the extent of the huge change mankind has wrought on the environment by cultivation, draining, planting and exotic plant and animal introductions. Having wrecked the natural environment for our own purposes, are we justified (or logical) in doing nothing when our woodland microcosm runs out of kilter? Or by such interference is the manager going beyond his right, acting like the Valkyrie, to choose who should be slain in the ecological battle?

FURTHER READING

Ferris-Kaan, R., Lonsdale, D. and Winter, T. (1993), *The Conservation Management of Deadwood in Forests*, Research Information Note 24, Forestry Commission, Edinburgh

Forestry Commission (1991), *Edge Management in Woodlands* (ed. R. Ferris-Kaan), Occasional Paper 28, Forestry Commission, Edinburgh

Forestry Commission (1994), *Wildlife Rangers Handbook* (ed. G.D. Springthorpe and N.G. Myhill), Forestry Commission Handbook 10, Forestry Commission, Edinburgh

Mayle, B.A. (1990). *Habitat Management for Woodland Bats*, Research Information Note 165, Forestry Commission, Edinburgh

Moss, R. and Picozzi, N. (1994), *Management of Forests for Capercaillie in Scotland*, Forestry Commission Bulletin 13, Edinburgh

Petty, S.J. (1989), *Goshawks: Their Status, Requirements and Management*, Forestry Commission Bulletin 81, HMSO, London

Robertson, P.A. (1992), *Woodland Management for Pheasants*, Forestry Commission Bulletin 106, HMSO, London

Rodwell, J. and Patterson, G. (1994), *Creating New Native Woodlands*, Forestry Commission Bulletin 112, Edinburgh

Woods for People   11

The kinds of municipal or communal forests which are common in most European countries are almost absent in Britain, and the country is poorer because of it. Among the benefits they provide is an awareness among townspeople of forestry matters, and an understanding of the realities of managing renewable resources; people take pride in their town's woods and they derive pleasure in using them for recreation. Many of these communal forests are extremely well managed and self-financing assets, and there is national benefit in having an electorate and politicians who, as a result, know about sustained management of natural resources in practice and not merely as catchwords.

An important government initiative was announced in 1996 to encourage more community ownership and participation in the management of woodlands in Great Britain. The scheme seeks to enable local community groups to buy forest areas and to develop partnerships between them and Forest Enterprise, the agency which manages commercial forest in national ownership. The Forestry Commission procedures for selling national forest areas into private ownership have been altered to give preferential treatment to communities who want to buy local woodland.

Where development agencies agree that acquiring an interest in woodland would further social and economic development, communities are allowed to buy the woodlands at the district valuation price rather than having to bid on the open market in competition with commercial buyers.

The gathering of funds needed for outright purchase of local woodland could be a considerable disincentive for many communities, particularly since they are likely to be uncertain of the commercial viability of the business. An important further aspect of the initiative, therefore, is the partnership scheme which allows communities to exert strong influence on the management of local woods which are still in the ownership of Forest Enterprise. In effect, the partnership could provide a community with most of the benefits of forest ownership with no financial risk whatsoever. The intention is to involve local people in the management of their neighbourhood forests and thereby to inspire new initiatives, particularly in respect of recreation, tourism and local employment.

These are important and welcome changes of national policy in respect of woodlands, and they go far to rectify the obvious lack of community and municipal involvement in forestry in Britain.

A previous move towards this objective was the offer of Forestry Commission grants for 'community woodlands'. Somewhat confusingly the grant is for the creation of woodland on private land, which is made open to access by people from a nearby town. The schemes have usually been at the initiative of the owner and many are some distance from the community they are supposed to serve, perhaps accessible only by car. Much more significant in most people's minds is the creation of woodland by a community, or by the landowner after real and close consultation with local residents.

Truly community woodland may arise from a variety of initiatives; typically a local resident, or a small group of local people, is the driving force. There is advantage to be gained if the driving initiative comes from a private person, rather than from a councillor or an official of the local authority. The key to success in a community woodland project lies in attracting the enthusiasm and direct participation of residents; a project conceived by the local authority planning department is liable to be regarded ever after on a par with street lighting and garbage collection: useful and something the Council does with the rates. The active involvement of people in creating woods and then in determining their management is the process in which a community comes to 'own' a woodland, whether or not they have any legal tenure.

Community woodlands must be accessible to the public for recreation; with advantage they may be linked to existing paths and cycle routes and may form part of a 'green corridor' from town to countryside. But, vitally, 'accessible' means both open and handy; it is not enough for the area to be open to the public if it is miles away from the people who need it.

A substantial difficulty is that the people who have the community woodland idea seldom have the technical knowledge to design one or have access to sufficient finance themselves to implement it. The former problem can be overcome by a learning process, while the latter may require guile. Success depends on persuading influential groups of people – councillors, planning officials, community councils, special funds and trusts – to provide the necessary resources, perhaps even convincing them that the idea for a woodland is their own. It is not a solution, however, simply to provide a forestry expert to produce a design, because the woods will then be regarded by the local people as the expert's woods, not their own.

The idea of a woodland for the use of the community may be floated with or without the precise location being fixed. If there is a choice of land, there may be merit in the innovator not tying the project firmly to one place at the start but allowing that important issue to be open to community discussion. If there is one obvious site, however, then immediately linking it to the project of a community woodland makes it easier for people to imagine and debate its usefulness and design. In some districts there may be land which is particularly suitable for the creation of local woodland, such as reclaimed open-cast mining (which is much cheaper to turn to woodland use than to agriculture, especially if the latter would then be put into set-aside) or former land-fill sites which may not be used for domestic housing because of methane gas discharge. Although it is unwise at the early stages to go into detail about the design and establishment techniques, it is essential to be clear from the start what the objectives for the woodland are and the broad advantages to be sought from it.

The case for creating a community woodland usually revolves around some or all of the following issues:
- Shelter from wind, perhaps summer shade
- Environmental improvement and diversity, providing a direct contrast with the built environment
- Contact with nature, valued as a focus for teaching, both informal and at schools
- Traffic-free safe area, especially for young families; a peaceful area for quiet recreation and relief from stress
- Active recreation – walking, jogging, trim

course, etc. – perhaps combined with mini-pitches for team games

- Pride in the quality of their place for local people

There is no doubt that well-conceived and well-maintained woodland enhances the value of property in the district and greatly improves the opportunities for attracting inward investment to the area. At least some people in society will regard it as a powerful argument for community woodland that trees soak up CO_2 and thereby in some measure counter the global greenhouse effect. Timber production is unlikely to be a major consideration but the fact that trees can produce useful material should not be ignored; all community woods require to be maintained – paths, stream bridges, fences, gates – and the possibility of making the area at least partly self-financing deserves consideration. Many town forests in Continental Europe, although small in area, provide examples of combining great beauty with high timber production; Breda in the Netherlands, Esseval Tatre in France and Couvet in Switzerland are among many.

It may be to the considerable financial benefit of local authorities to encourage the development of genuine community woodlands. Creating woodlands by effective planting to forestry standards can be contrasted with the intensive planting created by landscape designers, for instance around hypermarket developments and peri-urban roads. Whereas the latter may cost upwards of a quarter of a million pounds per hectare, forestry-standard work would be about one-hundredth of that sum, with better survival and lower maintenance costs.

Early in the creation of a community woodland it is essential to decide any special objectives the woodland is to serve and to win community backing for the concept.

Technical detail at this stage is best avoided. A person or group ('the innovator') should consider which local people might be specially interested, such as youth organisations, schools, the community council and local members of wildlife societies. Informal discussions should clarify the benefits each group would hope to gain from the project, information which will help with the next steps and will later guide the design of the facilities required by different users.

Development of the design should grow out of the previous discussion, reflecting the visions of the people who live nearby, probably both by asking them directly and by observing how they use the area. Activities such as football, riding bikes, walking dogs, pushing prams and many others are relevant, consideration being given to whether tree-planting might enhance them or interfere and whether zoning of activities would be necessary or helpful. In most instances, activities will take precedence over pretty views. In the process of developing the concept towards a plan, there is much to be said in favour of the group of local innovators having an 'away-day', arranging to visit one or more examples of community woodlands in being and, if possible, to meet members of the community group involved in their creation. The best consultancy is to meet one's opposite numbers, thereby gaining confidence, learning from their mistakes and hearing how they went about the task.

There are official bodies which can help: the Forestry Authority, the local council, the country's conservation agency, the Countryside Commission in England, local countryside trusts and regional woodland groups. All can give ideas and advice; some give grants.

It is at this stage of the design that arboricultural and silvicultural knowledge is required to enable the community group to interpret ideas into feasible plans. The design must make full use of any existing trees and reflect the risks of later windblow, leaf-fall, etc. It is obviously undesirable to give cover to glue-sniffers and the nervousness of people walking in close cover must also be addressed; planting should generally be kept back from path edges.

In the design process there is a choice of approaches: to produce a detailed plan, attempting to meet the community's needs at the first planting, or to move quickly to plant something 'cheap and cheerful' as a pioneer crop which can then be cut into as people's needs emerge. The former requires people to envisage from maps and on bare ground what the area will look like when the trees have grown, which is not easy. On the other hand, it means that no planting need later be 'undone' (perhaps a disheartening process) in order to adapt the quick planting to people's growing understanding of what they really need. The second approach means that the design does not have to be agreed before any planting is done; work can begin with cheap plants so that the low experience of the community members does not carry high costs and, a considerable advantage, work can begin while enthusiasm is high. Long delay while the final design is agreed may involve some loss of interest among people whose commitment is essential for success but who, once persuaded to become involved, 'want to get on with it'. There is a middle way, probably the best, in which only key features of the final design are agreed, with flexibility retained in respect of other areas, which are blanket-planted with a pioneer species.

New community woodland is almost certain to comprise both open and wooded areas, providing space for team games and wildflower meadows, if space allows, as well as good footpaths. In the final effect, wide-spaced trees rather than a closed woodland are likely to be appropriate. Since winter shelter and winter greenery are often desirable, the design should probably not be restricted to native species.

Do not overlook archaeology; important sites will be left unplanted and it may be possible, with professional guidance, to incorporate archaeological features as focal points in the woodland design.

The design process requires patience. It needs the services of someone with knowledge of trees and their growth characteristics, but it is important that the woodland consultant does not take too strong a lead role; in doing so, he could forfeit the enthusiasm of local people. Inconvenient though it may be, the task of the experienced person is to listen as the 'brief' emerges, to point out the implications and merits of various proposals and then to produce draft plans for the local people's amendment and approval. Only then may planting begin.

Moving from the first idea to the final planting of a finished design may take several years. In a specific example, villagers decided to turn a piece of unused ground adjacent to a play-park into a woodland for the community. The initiator, a resident with no official position, was the driving force both in focusing local opinion, so that in the end there was virtual unanimity, and in enlisting the support of the councillor and local authority officials to provide local-authority land and help with the design. The work involved planting trees (mostly native broadleaves), creating wildflower meadows, repairing stone walls on the boundary, making paths and building a footbridge. A high proportion of the local people turned out for the physical work, local enterprise trainees with a skilled supervisor undertook the wall repairs and bridge building, and members of a conservation trust volunteered to show beginners among the locals how to plant trees and to ensure safe working. The process, from the first concept, through 'getting all the donkeys to face in the same direction', to the field work, took four years. This may seem a long time, but the result is a woodland asset of which local people are proud and which has good prospects of being well cared for in the future because the local people are committed to it.

A vital stage is the expansion from the acceptance of the planting plan by the organising group to 'selling' it to the full community. Use should be made of school promotion, local radio, local press articles, maps, drawings, photo-montage and models in the local library, all to show how the woods should fit in with cultural, heritage and archaeological features, and with neighbouring land uses. A public meeting on site may be more effective than one indoors. The clear objective must be to engage the enthusiasm of local people and get them into the field to do part of the work themselves, whether planting trees or pouring out tea or lemonade for those who are planting.

Leaders of groups engaged in any conservation work, including forestry, should be sure to have proper insurance cover for any personal injury that may arise to volunteers. The provision of insurance and training are two great advantages of organising a project under the aegis of one of the national bodies for conservation volunteers (see addresses in Appendix).

The involvement of conservation trusts and local enterprise agencies should be sought and local commercial firms approached both for financial contributions and to enquire if the community woodland project could be fitted in with the management training scheme which most businesses use for staff development; where a firm's personnel go,

their financial support often follows. It can scarcely be over-emphasised that community woodland work requires a long-term commitment; it is an essential function of the driving group to ensure that enthusiasm and commitment are sustained. The availability of funds from the Millennium Commission and its associated trusts now provides a massive encouragement for community woodlands.

Common to almost all community woodland projects is the involvement of young people, particularly school groups. School education, both curriculum-based and for general development, has much to gain from access to traffic-free, natural-environment open space. Equally, engaging the interest and personal involvement of young people is the surest way, perhaps the only way, of ensuring the sustained management of community woodland and, in the short term, guarding against thoughtless vandalism of it. Winning the support of local schoolteachers is an obvious and early step for the organisers.

There is no single recipe for creating community woodlands; the design must reflect the character of the site and what the local people are looking for. Our society's current ideas of beauty in landscape are not unchanging. Not so long ago, the formal walk, the straight avenue, the artificial lake and carefully placed folly were the delight of people who saw ideal beauty in trim cultivation and only ugliness in wild mountain and moorland. The concept of wilderness as desirable and beyond price is of relatively recent origin in Britain. Attitudes have changed remarkably over the last 150 years, expressed in a general disenchantment with city life, a growing concern at the pollution of our environment, aversion from whatever is mass-produced and an appreciation of things unique and primitive.

Against this background, trees and

woodland are seen by very many people as highly desirable, particularly when they lose the obvious features of a plantation. They are not mass-produced or ephemeral but are associated with naturalness, offering an escape from traffic and the pressure of the built environment. The woodland's area, structure and composition deeply affect its capacity to 'carry' people, in the sense of satisfying their needs and accommodating their activities.

It is intriguing that virtually all the exotic trees grown in Britain, broadleaves and conifers, including such important industrial species as Douglas fir and Sitka spruce, were introduced without regard to their potential for timber-growing, but simply for the adornment of the parks of wealthy landowners, as an expensive form of nineteenth-century one-up-manship which has benefited far more than those who paid the botanical explorers and the tree-planters. The fashion for creating arboreta has greatly enriched the country scenically. Apart from the rarity value of a species in the years immediately after its discovery and its introduction to gardens, what qualities make trees attractive to people?

Large size and old age in trees are seen as desirable attributes. (The readiness of the uninitiated to overestimate the age of a tree is matched only by the eagerness of the grower to exaggerate its girth and height.) Certainly size and age appear to attract more attention than perfection of form, line or proportion. People wonder at a craggy old larch, a hoary oak or a massive wellingtonia. In woods for people there is a strong case for conserving big trees, which involves retaining some trees beyond their normal 'sell-by date'. People's enjoyment is enhanced by a woodland which contains some big trees. (Foresters are well aware of visitors' illogicality; it is not uncommon for wondering admiration of magnificent 150-year-old Douglas fir to be accompanied by regret at 'the dark conifers' nearby, conifers which are in fact young Douglas fir – the only way we know of producing the big ones!)

A quality which people generally value among trees is variety: variety of colours, shapes, texture and species, and varieties of size and age, which provide the best context in which to show the largest and oldest. We find pleasure in making visual contrasts and comparisons, and are quickly tired and irritated when the eye finds no dissimilarity to arrest it. In planning for variety, however, it is a mistake to make every tree different. This would be chaotic; a few well-placed contrasts suffice: one copper beech, one spiky grey Atlas cedar against silver birches, a clump of scarlet rowans or geans against evergreen conifers. The differences of colour presented by various species at different seasons of the year are important.

One of woodland's outstanding qualities is its capacity to provide privacy; its people-carrying capacity is orders of magnitude greater than the seaside or moorland. Whilst people who are naturally gregarious delight in forming part of a crowd, others seek to relax in small groups. For the latter, the open beach or moorland quickly becomes unacceptably crowded with a few others in sight, whereas woodland occupied by far more people gives them all the illusion of having the place to themselves. The avoidance of long, straight paths and the provision of some areas or strips of understorey to reduce the awareness of other people in the wood are therefore sensible design points.

Nevertheless the creation of a complete understorey of shrubs in a wood for people is unhelpful. Paths which run as long corridors through dense shrubbery make people nervous lest they be pounced upon and, in any event, are uninteresting because they afford no long views. Openings which allow a good view out of the woodland provide pleasurable surprise, as are vistas within the wood across open grassy areas and interesting ground vegetation which may be blaeberries or bluebells. Structural diversity is desirable in woods for people; group clear-felling or group-selection systems are especially appropriate because they ensure the woodland is both continually sustained and never uniform, ecologically or visually.

On the whole, people dislike changes in their environment (even what is generally held to be improvement), and they especially dislike sudden change. In a wood for people, it is easy to predict that clear-felling and replanting a large area will produce a storm of protest. There is great merit, therefore, in the following advice:

- Tell people what management operations are envisaged, even in privately owned woodland where people have the privilege of access, so that they are not caught by surprise by the changes, so that they may understand the ecological necessity of tending and renewal, and so that their ideas may be incorporated into the plans.
- Arrange that the items of change are individually small and occur continually, a

kind of management by stealth for woodland heavily used by the public. In this respect the group-selection woodland holds particular interest – it is a steady-state woodland, silviculturally ever-changing but, to the visitor, apparently unchanged.

Woods for people need not be unproductive and it is a mistake (because it is a missed opportunity of informing visitors about real woodlands) if one pretends that operations such as tree-felling and planting do not go on. People must be kept at a distance when trees are being cut and machines are working, but experience has shown visitors' deep interest in woodland work. They want to be reassured that regeneration is to accompany cutting; there is an opportunity to say in a notice what is going on, what markets the produce will go to and what the new tree stand will comprise. Why is this so seldom done?

Except in large forests, public access into the woods by car is not acceptable, so provision should be made for cars to be parked at the edge. If the scale and budget allow, the car park will be the place to locate lavatories and other built facilities, such as an

REGENERATION IN PROGRESS
A succinct notice at regeneration areas in France, to reassure visitors this is sustained working, not exploitation.

information board about walks and forest operations. Here too will be a suitable place for picnic tables.

In larger woods it may be appropriate to divide the area into zones for different activities, particularly to separate the quiet and informal ones from the formal and the noisy, such as ball games, with the aim of reserving some parts for minimal disturbance. In small woods it is better to avoid the formal and noisy activities and to concentrate on quiet recreation.

In all sizes of woodland used for recreation, full advantage should be taken of features such as rocky bluffs, ponds and streams to enhance the variety, and way-marked paths should lead the visitor to them. It may or may not be feasible to erect a hide for bird-watching at a pond or to locate picnic tables there. In all but the smallest woods, way-marked paths are helpful; most townspeople have serious anxieties about becoming lost in woodland and not finding their way back to their car, perhaps the result of early reading of fairy stories in which the main purposes of forest were apparently to hide the big bad wolf and to allow children to lose their way. Colour-coded paths, marked at intervals and at all junctions by painted stakes, give reassurance. In larger woods they should be reinforced with leaflet maps and by notices at the start showing distances and average times for the circuit. The planner should assume that the path will be used by less-mobile people and only if proved unavoidable should the exception be allowed.

Woods for people should be regularly inspected for hazards, including dangerous trees, especially near paths and play areas. The dates and nature of these inspections and consequent actions should be recorded in the woodland management plan. Rubbish, graffiti and the results of vandalism should be cleared up promptly; failure to do so creates an impression that these are acceptable, which encourages more of the same.

On the urban fringes, some woods are used regularly for dog-walking, and one of the most difficult problems may be fouling of paths. This is usually most severe at the entrance to the wood; owners bring their dogs by car to the car park and lead them on to the path where, almost immediately and by habit, the dogs defecate. An alleviation, although probably never a solution, may be achieved by having two entrances to the path network and persuading people who regularly bring their dogs to lead them on the 'dog-bypass' which should be so marked; a short loop of 40–50m may be sufficient to concentrate a lot of the fouling where it is much less unacceptable, leaving the main entrance cleaner.

In larger woods for people (and even in some smaller ones) the interaction of different traffic may have to be relieved by zoning routes. Some mountain-bikers are aggressive, some horse-riders are insufficiently considerate towards walkers and some walkers are not tolerant of horses. If woods are being designed for a high level of use by people it is sensible to think of separate walking paths, bridle paths and cycle tracks. None of these should be available to motorcycles, for which provision is best made elsewhere.

The value of community woodlands around towns is perhaps highest when they both serve the immediate needs of local people for recreation and form, like a string of pearls, the focal points on a communication network for walking, cycling and horse-riding, across the countryside between towns. The strategic design of the network is the responsibility of local authorities and countryside agencies; this chapter is aimed at giving ideas to those who are concerned with the woodlands which may be the individual pearls.

Woodland for people may enhance the

success of commercial enterprises nearby, firewood sales, greenery and Christmas trees, bicycle-hiring, barbecues or cottages. The production of all these may be compatible with management for people. Indeed, it is usually an error of judgement, except in the very small properties, to divorce the management of woodland for people from woodland for timber production or for wildlife. Unless woodland for people is managed sustainably, it will evolve into unsuitability and later into senility, as so many existing parks have done. As in all woodland, there is danger in allowing management to freeze into preservation, where conservative management is needed. Woodlands require maintenance, for which the money may come from user fees, a public or private subsidy or produce sales. Most effective have been the instances where good conservation and public access have been carried on the back of timber sales in genuine multi-purpose practice.

The availability of land may be critical. Where there is land in the ownership of the local authority, as there may be after industrial dereliction, the completion of public housing schemes etc., this is the obvious starting point for a new community woodland. Otherwise, if land has to be purchased, there will be a hard financial choice. Land on the fringe of a town or village carries 'hope-value', the owner's hope that planning consent for built development will escalate its market price. A community woodland far from the town might cost less but would be of far lower value to people than one much smaller but close to town. The people who need the woodland most – the school, mothers pushing prams, young children needing a traffic-free area – are not sufficiently mobile to benefit from a wood even a moderate distance away.

Where there is any choice in the matter, careful thought should be given to the continuing ownership of the land. It may appear that the local authority is the obvious custodian of the title. It should not be forgotten, however, that some local authorities (town councils in Scotland, and perhaps elsewhere too), when faced with a merger with a neighbouring or new council, have sold the town's woods to private developers for destruction in order to avoid passing them to a rival council, in spite of the fact they were assets held on behalf of the people who continued to live there and whose needs were unchanged. An alternative may be for the land title to be held by a charitable trust, the trustees being the current office-bearers of various named posts such as the head teacher of the local school, the president of the tenants' association, the chairperson of the community council and so on. In all circumstances the powers of the trustees to dispose of the assets should be fully detailed and it should be stated that, in the event of the local trust becoming inoperative, ownership would revert automatically to the local authority or, perhaps, to a named national trust so that the community's woodland would not be negligently or wilfully lost.

The proprietorship of community woodlands is a key feature of the 1996 initiative for ownership or partnership with Forest Enterprise and it is relevant to the Forestry Commission's Community Woodlands grants. It is possible, by monetary compensation in the latter scheme, to obtain public benefits from privately owned woodland on urban fringes, but the potential conflict between management for public advantage and private gain is always present. The possibility of such conflict should not be ignored in a woodland system where continuity of purpose and practice must be sustained over many decades. Provided that there are safeguards against the kind of injudicious disposal which occurred in the earlier reorganisation of Scottish local government, there is considerable advantage in woodlands on the fringes of cities, towns and villages being held in local-authority proprietorship. Such woodlands are public assets which generally hold or increase their capital value over long periods and yield great benefit in public education and pleasure, and potentially in local employment. The fact that this is unusual territory for British public administration should not be a deterrent; Continental Europe has excellent examples for every size of community: the Forêt de Soignes' superb beechwoods at Brussels; Amsterdam-bos, created on wasteland in the 1930s; the city forests of Frankfurt and Cologne; down to the public woodlands of small towns and villages. Most are well managed and yield great amenity benefits, some substantial money income. It is to be hoped that the new scheme in Britain to encourage community ownership and management will succeed in achieving similar results.

FURTHER READING

Countryside Commission (1990), *Advice Manual for the Preparation of a Community Forest Plan*, Countryside Commission, 19–23 Albert Road, Manchester

Forestry Commission (1991), *Community Woodland Design Guidelines*, HMSO, London

Forestry Commission (1996), *Involving Communities in Forestry*, Forest Practice Guide 10, Forestry Commission, Edinburgh

Hibberd, B.G. (ed.) (1989), *Urban Forestry Practice*, Handbook 5, Forestry Commission and Department of the Environment, HMSO, London

Hodge, S.J. (1995), *Creating and Managing Woodlands Around Towns*, Handbook 11, Forestry Commission, Edinburgh

Farm Woodlands

When farmlands are exposed to high winds, interspersing them with strips and masses of plantation is attended with obviously important advantages. Not only are such lands rendered more congenial to the growth of grass, and corn, and the health of pasturing animals, but the local climate is improved.

Walter Nicol
The Planter's Kalendar, 1812

There are all kinds of farm woods: excellent, useless, nothing much at all, eyesores, delights.

A few years ago, a farmer in the Scottish Borders, made a crucial comment to me, and might have been speaking for hundreds of fellow-farmers: 'I wish I had thought harder before planting these trees 16 years ago; it's 16 years wasted.' He had thought some small plots of trees would be a good addition to his farm, and a chance meeting with an employee of a private forestry company had resulted in the planting of three small blocks of Sitka spruce. 'It seemed a good idea at the time because they were cheap and they were the same trees the company was using for the big pension-fund planting at the head of the valley.' Although the trees had grown strongly, he now had severe doubts and I had to agree with him; although Sitka spruce can be a super tree in extensive production forest, he had little hope of selling the produce from his tiny blocks at any profit, and, in any event, that was not what he had had in mind when he put the trees in the ground.

John Dixon, on the other side of the border, knew pretty well what he was aiming at when he began planting a woodland in 1959. He had several objectives: to convert spare time into an asset or, better still, into hard cash; to provide a nest-egg for one of his sons, since he thought all three could not make their living on this farm; to improve the shooting; and to give some shelter, particularly some wooded grazing where animals could be held and hand-fed in hard weather – 'to keep the frost off their backs', he said. Not least in John's mind was that the piece of land grew bracken so tall that he could lose his cows in it, which was inconvenient. His objectives for the woodland were joint production and protection, and so they remain.

This upland dairy farm of about 80ha carried a stock of 70 pedigree Friesian cows in-milk, with young stock, and a small flock of 40 Greyface ewes which were put to Suffolk tups. The winter keep for the cows was mostly silage made on the farm, with hay and concentrates. All the work was done by the farmer, his wife and their family of three sons. 'Spare time' might seem to be a scarce commodity on a family farm with 70 milk cows, their followers and 40 sheep, but that is what John found for his remarkable woodland enterprise.

The area John chose for his planting was 10ha of land which could not be grazed effectively and, because of its bracken, was more of an embarrassment than an asset. It adjoined an existing group of larch which now became part of a larger wood. Work began in 1959, when John planted 3.6ha, and continued each year until 1966 in areas between 0.4 and 2.0ha. A great variety of species were planted, principally hybrid larch, Sitka spruce and Scots pine, with some western hemlock, grand fir, beech, sycamore, maples, rowan and copper beech. Some were bought from nurserymen and some raised on the farm itself. John did the planting himself 'between milkings' and his sons did the weeding – no light task on this land with heavy bracken. The success was virtually 100 per cent and on this fertile land the trees grew particularly well.

The spreading of the establishment over several years was crucial, avoiding the peaks of demand for maintenance which simply could not have been met on a family farm. Eight years of steady planting were followed by seven years of early tending, none beyond the scope of the family. Then, at age 15, the oldest planting was ready for its first thinning, beginning with the hybrid larch. Early selective thinnings are virtually an impossibility for the manager of a large commercial forest, bound to lose money and now largely avoided in 'big-time' forestry. They are, however, a boon and a blessing to the farm-forester if he can carry them out, providing an early return on his investment, although their main objective is to improve the crop which remains.

In winter, when the cows were indoors, John and his sons cut the thinnings, again between milkings, pulling the long stems down to the steading with the farm tractor. There the trees were cut into hedging stakes, fence posts and the like, to be taken to town every market day, with firewood for home use as a by-product. At market people knew they could always rely on being able to buy from John Dixon the stakes, posts and rails which every farm needs continually for repairs. So the business was built up. In the 1970s and 1980s, as now, this income was tax-free; John was turning time into money, at no cost whatever to his stock-farming enterprises.

In addition to the produce sales, the farm benefited from shelter for the stock, most of all for lambing, for early-spring grazing and for cover in which to hold the sheep in severe weather, where they might be fed. The woodland held game birds; pheasants were reared and the shooting greatly improved.

In his own locality John Dixon was a pioneer of this kind of farm-forestry, a man with vision, persistence and great enthusiasm. He created a living woodland which has been successful both in production and in its good appearance. A major contribution was his demonstration of how woodland work could be integrated with animal management on a busy family farm. John's original intention in planting trees was to create an appreciating asset which his sons might realise when the trees were mature, but from year two he envisaged sustained management: the need to restrict planting to what could be fitted in with the regular farm work, and the consequent advantage yielded by that spread of ages in the trees as they came to usable size. Sustained inputs are linked to sustained yields. Nobody taught him; he conceived sustained-yield management for himself.

For a very long time British farmers have differed from their Scandinavian and Western

European counterparts, many of whom own woodland, manage it effectively and regard it as a normal income-earning asset on the farm. Traditionally the British farmer has not been his own forester, nor has he been actively interested in woodlands, but recently the position has shown some change. There are two distinct approaches to these land-use decisions. One – historically common – is for forestry and farming to be regarded as directly competing systems which can 'win' and 'lose'. The other approach allows the possibility of an integration in which the enterprises may benefit from the interaction between them, so that the total output from the integrated system is greater, at least in the long term, than that which could be obtained from the same resources separately, either in farming or in woodland management. The second was John Dixon's approach, and he applied it well on his farm.

Very recently I went back to John Dixon's wood. John died several years ago and his sons are farming now. I walked through the wood again, admiring what I saw and full of wonder at what was done and is still happening 'between milkings'. Whereas I remember light fencing stakes being thrown into a trailer behind a car to go to market, the produce is now heavy straining posts. The larch is sawn into fencing stobs and there are pine and Douglas fir sawlogs. John's eldest son Bryan said the trees grow so quickly and produce so much, it is hard to know where all the wood comes from. He takes pride in what his father did, and the late John would certainly be pleased at the way his wood is managed now.

One of the country's most successful big-estate land managers, talking about farm-forestry, shrewdly remarked that the way to make money from it was to do the work yourself and pocket the grants: 'Farmers don't make money by hiring contractors to do the milking or saving the hay but by doing the work themselves and with the farm staff; the same applies to woodlands – they should do it themselves.' That was certainly borne out by John Dixon. In effect, forestry is a means of converting time into capital, working on a scale and to a schedule which fits in with the farm work.

Recent studies of the potential for extending woodlands in the lowlands of England and Wales and also in the Grampian region of Scotland have confirmed the financial benefit of using farm staff for the work (see B. Bunce *et al.* 1994 and R. Lorrain-Smith *et al.* 1997). These studies compared average gross margin returns when land was transferred from existing farm crops and grass to woodland, making a variety of assumptions about the species of trees planted. The results consistently showed the cash advantage of the forestry work being done by farm staff rather than by contract. Equally important is the integration of the woodlands with other farm work. A farmer should not operate at a scale and speed beyond the capacity of his staff; contractor-scale planting will probably have to be followed by contractor-scale everything else, and the woodland will never be integrated.

There is no single recipe for successful farm woods; each farm and each wood has to be considered afresh, but there are certain principles which should be respected.

Above all, the farmer must decide the 'job specification' for a wood before starting work; otherwise he is liable to end up with something that is an embarrassment and a liability. Creating a woodland is a long-term undertaking which deserves careful thought to get it right. Initially there is a triangle of influences and pressures: the objectives the farmer wants to achieve; the species which are intrinsically well suited to achieving those objectives; and the site factors limiting the range of trees which realistically will grow well on that land.

The purpose of the woodland may actually determine where the wood has to be located. A shelter strip may have to be along the edge of a particular pasture; a group of trees to screen a building will have to be placed right there. In these instances the site factors (soil, exposure, etc.) are set and the choice of species available to serve the objectives will follow accordingly, restricted on inhospitable sites, wide on easy ones.

In other circumstances, the land for woodland planting may not be so closely fixed and the farmer can use his own discretion in balancing locations (each with its limitations of soil and exposure) against the objectives he is trying to achieve and the tree species which would help him to do so. An initial selection of the location with the poorest soil and most severe exposure might restrict the species for planting to those which would not meet the farmer's objectives, so a series of alternatives may have to be considered, trading off less convenient locations for a wider choice of species to attain the objectives.

The farmer who has an eye on timber production should be clear that the forest industry provides great economies of scale. The big producer who can offer a timber lot of 10,000 tonnes, of one or two species, to be machine-harvested by a massive processor and carried away in 38-tonne loads will always leave the small grower at a huge disadvantage if they are operating in the same market. Nevertheless, there is no need for the farmer-forester to be discouraged, because he can do things the big producer cannot do: early selective thinning, using nurse crops, pruning, developing special markets. The big producer will be confined, more or less, to relatively difficult upland sites and a low-input system, generally growing Sitka spruce and low-grade pine for bulk markets. Farmers should try to avoid direct competition with such growers; instead try to find specialised or local markets, including those for spruce and pine if these are to be grown, and generally try to aim for quality production rather than low grades.

High-quality wood production is generally easier on fertile sites than infertile ones, and it is certainly easier in sheltered areas rather than exposed ones. On fertile sites, the small grower should plant demanding tree species which will produce intrinsically high-value material, rather than simply growing a larger volume of low-value wood, which is inevitably the option taken in the extensive plantation and has been, unfortunately, all too often the ready-made advice from those with experience of 'big-time' forestry. Thus it is more appropriate for the small grower to grow a demanding species such as ash, sycamore, gean or Douglas fir than it is for a bulk producer in the now-traditional low-input system.

More especially, however, the small grower should plan for harvesting a series of products over the years, and particularly for early harvests. Regrettably, the manager of extensive low-input plantations may now see his ideal as planting a single-species crop which will require no tending whatever (and no product to sell) until it is machine-harvested at age 35, when the whole process may be repeated. In sharp contrast, a high-input system with mixed species may be operated to yield a first harvest at age eight, then hedging stakes of hybrid larch at age 15 (like John Dixon's), fence posts from 20 to 30 and so on, while the really valuable and interesting material is continuing to grow.

Aiming up-market with high-value products, or at least versatility in marketing, especially in serving local and specialised needs, should be the watchword for farm-forestry (preferably having done the planting in-house and having banked the planting grants).

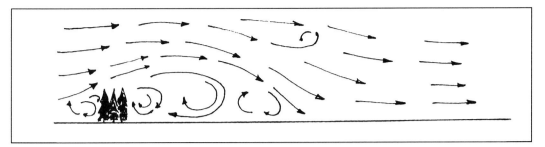

WINDFLOW OVER A SOLID BARRIER
Windflow over an impermeable shelterbelt. Large standing eddies develop just behind the belt, giving a very short sheltered zone.

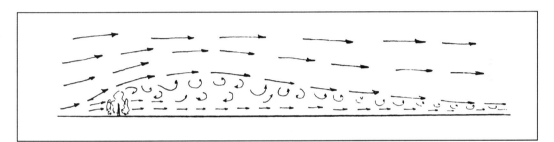

WINDFLOW OVER AND THROUGH A PERMEABLE BELT
Small rolling eddies hold the main flow well above the ground and give shelter to 20 belt heights and more downwind.

FARM WOODS FOR SHELTER

The planting of trees to provide shelter calls for special thought. There is undoubtedly a temptation on the farm to plant a strip, call it a shelterbelt and try to make it serve every purpose from a hard standing for winter feeding of cattle to growing future veneer timber.

If the farmer wants above all to give extensive shelter to a field or a wide area of grazing, then by all means he should plant a shelterbelt, but he should consider carefully the problems of making it serve other purposes.

- A long, narrow strip is an expensive shape to fence and difficult to maintain or regenerate.
- In a place exposed to wind, edge trees suffer such stress that the achievement of high-

quality wood is virtually impossible from the margins.

The provision of extensive shelter over whole fields (as distinct from storm shelter for animals or shelter for a house) requires that the wind should be slowed down and filtered, not that a barrier is put up to stop it.

When the wind meets a solid barrier, like a wall or a thick belt of trees, the whole windstream is forced over the obstruction and immediately plunges down on the lee side. In fact, a 'standing eddy' develops on the lee side, with the strong flow near the ground towards the barrier. The zone of shelter is very narrow, immediately on the lee side. A solid barrier is of no value for providing extensive shelter to a field. Indeed, it may be worse than useless if these are in arable crops; often we see cereal crops laid by the standing eddy immediately in the lee of a wood or strip of dense trees.

The narrow zone of intense shelter may be just what is wanted to give storm shelter to stock, providing a place where stock can be held for convenient feeding, perhaps for lambing.

The design secret for true shelterbelts is that only some of the windstream should be made to climb over the trees, while some, at least 40 per cent, is filtered through. The air passing through both slows down and forms a large number of small rolling eddies, which have the effect of holding the main windstream above the ground for a considerable distance downwind.

There is a close connection between the height of a shelterbelt and the width of the downwind sheltered zone: the taller the belt, the greater the downwind shelter. A permeable shelterbelt reduces the wind speed at 1–1.5m above the ground to some extent for about nine times its height upwind and about 30 times its height downwind, but really useful shelter should not be expected more than about three heights upwind and 20 heights downwind. This means that a well-designed permeable shelterbelt of trees 18m tall (60ft) should give useful shelter to about 54m upwind and about 360m downwind,

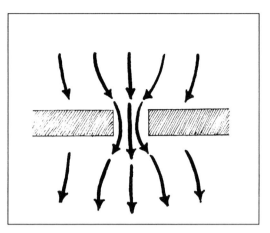

WIND ACCELERATION THROUGH A GAP
Wind 'crowds' through, making a gap a very difficult place for replacement trees.

assuming that the wind is blowing at right angles to the axis of the strip; if the wind strikes the belt obliquely, the protected zones are reduced.

From this it follows that permeability is an extremely desirable design feature for effective shelterbelts. Ideally there should be a large number of small openings throughout the height of the belt, allowing 40 per cent of the wind through. It is especially harmful if a shelterbelt is 'porous' merely because the bottom has become open and draughty, as happens when belts become older and are poorly maintained. In belts where the bole space is open, the wind accelerates through the belt, actually travelling faster than the free wind speed, to the discomfort of people and stock. Such decrepit belts are worse than useless.

The replacement of narrow strips of trees always presents difficulties, not least with those which have been created explicitly to provide shelter, since their removal means the area is deprived of shelter until the trees regrow.

When a narrow strip of trees becomes draughty below, the accelerated airstream

DRAUGHTY BELT
A 'shelterbelt' only in name. In fact the wind through the bole space is accelerated, not reduced.

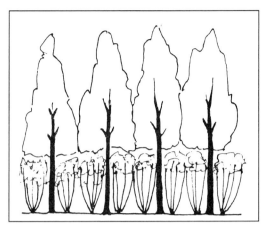

STANDARDS AND COPPICE BELT
A design to suit very narrow belts, e.g. poplar with lime coppice. Thin the coppice to regulate permeability.

makes conditions especially difficult for plants, so that underplanting followed by progressive cutting of the old trees is seldom successful. Cutting out a short section of a strip creates, in any windy situation, conditions which are virtually impossible for trees; through a narrow gap, the airstream may be twice the free wind speed.

One suggestion made in literature but usually impractical in Britain is to plant the replacement strip alongside the existing one, which can be removed when the replacement is well established. Clearly this requires a temporary doubling of the area under trees, which may not be acceptable. Another suggestion is to cut the trees on half the width of the strip and replant there, followed by the same operations on the second half when the first is well grown. While this may be feasible in broad strips, it is seldom feasible in narrow ones. In a narrow strip of more or less uniform structure, complete removal and replanting may be the best procedure for regeneration.

One of the great merits of a cellular or group-structured belt is that it may be possible to cut and replace a single group (by planting or by coppicing) and thereby manage the woodland on a perpetual basis. In effect, it is managed on a group-selection basis, with the important proviso that adjacent groups must be opened at such intervals that through gaps are not created.

In managing narrow strips, the usefulness of coppice should not be overlooked. Some excellent results can be obtained by planting at wide spacing the trees which are to provide height, with the lower shelter provided by coppice (sycamore, hazel, lime, etc.), the stems of which are thinned out to provide the desired permeability and which have the capacity for rapid regrowth when regeneration is required.

Shelter strips should not be allowed to become degenerate but should be tended to retain the lower cover before they become draughty, by group-planting or coppicing. Once a belt has become draughty it creates conditions which are usually worse than having no trees at all. Timely management should be the rule.

Special conditions may require the use of unusual species. One of the most successful specialist vegetable farmers in East Lothian, the late Mr John Dale, aimed to maintain 25 per cent of his farm area as shelterbelts. On sandy soil near the sea, many of the belts were planted with sea buckthorn, *Hippophae rhamnoides*, native in the district and very well adapted for the purpose. Although this shrub does not reach more than about 7m in height, the ease with which it can be propagated and its speed of growth make it useful for shelterbelts in sandy soil, and John Dale's narrow strips at close intervals over the farm created excellent conditions for growing vegetables. When a belt became degenerate, he simply cut it over completely and allowed it to regrow.

Some design features, both desirable and

WIND PRUNING AND FLAG CROWNS
Shelter strips should contain their crowns within their own boundaries. The flag-sided crowns of beech are especially undesirable, long and casting heavy shade.

undesirable, have an important influence on the selection of trees for shelter planting, some glaringly obvious.

1. Trees should be naturally windfirm, deep-rooted on the chosen site and capable of forming a permeable belt (spruces are not good in this regard, since the dense, twiggy crowns are almost impenetrable by wind; beech casts such heavy shade it almost precludes having any shrub layer or understorey and is very liable to become draughty below).

2. Trees should be able to maintain their crown in exposure, e.g. sycamore, sessile oak, Turkey oak, whitebeam, grey poplar, noble fir. A poor performer is Douglas fir, the crown of which becomes tattered.

3. Since the sheltered zone is a function of the top height of the belt, tall trees are desirable for the core of the planting.

4. Trees which develop elongated 'flag' crowns

GROUP DESIGN FOR SHELTER
The core groups give height, which determines the length of the sheltered zone downwind. The marginal groups prevent the belt becoming draughty. The group design makes for simpler tending and continual renewal. Crowns should be held within the field boundaries. The hedge is useful in giving low shelter.

downwind are undesirable, because they overhang the boundary and adversely affect field crops and grazing. Beech is particularly bad in this respect.

5. For the core of the shelterbelt, long-lived trees are more valuable than short-lived (if the main product is shelter, 60 years of shelter from one planting investment is a better bargain than only 30 years).

6. If there is a risk (or an intention) that stock may have occasional access to the belt, trees which have shallow feeding roots (e.g. beech and spruce on some soils) or smooth, thin bark which stock will gnaw and strip (e.g. beech, spruce, sycamore, Norway maple, silver firs, lodgepole pine) are unsuitable; trees which develop a flaky bark at an early age (Scots pine, larches, sessile oak), which animals obviously find unpleasant on their lips, are more suitable.

Since almost all trees tend to develop clear lower boles as they become older, it is difficult to avoid a belt eventually becoming leggy and draughty if one species is planted across the full width. Consequently, it is generally helpful to design a belt with a central core of the tall, long-lived, most-windfirm species, with marginal trees flanking them to windward and leeward. The marginal trees can be thinned to adjust the permeability of the belt and, if they are planted in small groups, there is an opportunity to achieve a good variety of species and hence of micro-habitats for wildlife. Rowan, Swedish whitebeam, hazel, hawthorn, Midland thorn and field maple are among the most widely suitable marginal trees because they are naturally sturdy and of small stature; they provide interest in foliage and blossom, and support birds and other wildlife.

If possible, the layout of planting for shelter should avoid wet hollows, where tree rooting will be poor and windblow likely. Such

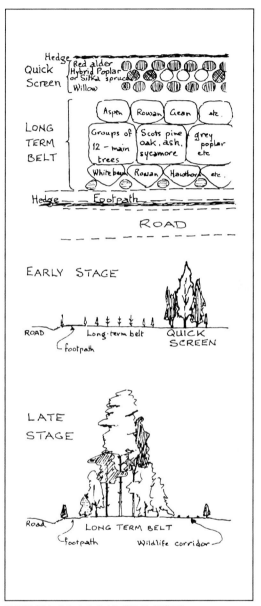

DESIGN OF A ROADSIDE STRIP
The design uses quick-growing, rather short-lived species while the long-term belt is growing. Later the early screen will be cut to provide an open corridor for wildlife on the lee side, away from the road and footpath.

windblow tends to spread inexorably and a corridor blown through a shelter strip causes intense funnelling of wind, destroying the very shelter which is being looked for. Frost-hollows should also be avoided; the designer should consider whether the belt itself, when it grows, may create frost-pockets which would be unhelpful to stock.

The advantage of longevity offered by some species may be offset by their slow development, and there is no doubt that most farmers, having made up their mind that shelter is desirable, want shelter quickly (if not yesterday). There are some species, like willows and poplars, which can grow very rapidly in the few years after planting, and it may be possible to plant the long-lived but slow-developing trees (say sessile oak) in a matrix of short-lived quick-developers which must be cut out early in order to allow the permanent trees to develop. This was well shown in a design for planting along a road where immediate screening was required for a land-fill site; after about 15 years it was expected the belt should provide shelter and a wildlife corridor.

The trees which are to form the core of a shelterbelt should have well-developed, strong crowns, which may be encouraged by wider than normal spacing of the trees at planting or by early thinning or respacing; above all, these trees should not be allowed to grow crowded and spindly. Respacing may be readily achieved by planting the core trees at rather wide spacing and interplanting with 'filler' trees, which may act as nurses (see chapter four) and are easily removed at age eight to 12. The nurses can serve double duty if they provide shelter in the early years and are themselves marketable, for instance as Christmas trees. The core trees should be chosen for robustness, longevity and height; the eventual performance of the belt in downwind shelter will be a direct function of its height.

Among the most useful trees for shelterbelt cores, according to local soil conditions, are the following (see chapter three for site preferences and limitations): sycamore, Norway maple, sessile oak, Turkey oak, common lime, Scots pine, noble fir, Leyland cypress, holm oak, sweet chestnut, horse chestnut, Corsican pine, ash, grey poplar.

For short-stature trees alongside the core (the retro-margins): whitebeam, Swedish whitebeam, rowan, gean, bird cherry, alders, aspen, Midland thorn, hazel, field maple.

For shrubs: holly, hazel, hawthorn, Midland thorn.

For hedges: hawthorn, Midland thorn, holly, beech, hornbeam.

On difficult sites some other species have particular applications:

On very exposed upland sites: noble fir (see Sitka spruce below).

On seaside sites with possible salt spray: Corsican pine, sea buckthorn, grey poplar, sycamore.

On exposed upland sites with moderate soil: red/grey alder, downy birch, goat willow and hybrid larch.

On exposed upland sites with infertile soils: lodgepole pine, Scots pine, Sitka spruce (depending on soil and moisture).

On many harsh sites Sitka spruce is very resistant to blast, especially when young, and it can provide quick shelter, but it is difficult to maintain in a narrow strip, its shallow rooting on many soils makes it rather prone to windblow later and its dense crown makes it difficult to achieve the desired degree of permeability; these comments apply speci-fically to narrow permeable belts. It is susceptible to periodic defoliation by aphids, especially in low-rainfall areas.

On reclaimed industrial sites: red alder (not long-lived), goat willow, downy birch.

For rapid temporary shelter and screens: white willows, some poplar clones (Balsam

Spire), red alder, common alder, cordate alder, Sitka spruce, Scots pine, Norway spruce (as Christmas trees), birch (although its whippiness makes it a difficult companion and nurse on windy sites).

Especially in the hills, farmers often want shelter to which stock may move in severe weather, not aiming to shelter a wide area of the grazing. For this purpose dense woodland blocks are very suitable, some having been planted as simple rectangles, some shaped to suit a land feature such as a gully dangerous to the stock, and some designed with a sinuous edge which allows animals to find shelter in winds from several directions: Manx-legs, Maltese crosses, L-shapes and crescents have all been used in the past. The narrow zone of deep shelter given by the impermeable trees has been described by some farmers as a positive advantage, since it encourages the animals to congregate where they may be easily found for feeding during a snowstorm. Nevertheless, the concentration of animals in wet weather may result in severe poaching of the land and heavy trampling, so the choice of site is important to make best use of firm ground as hard standing in the lee of these blocks.

Unfortunately in the past 50 years many of these storm-shelter blocks have been planted as simple rectangles of pure conifers. When set in an almost treeless upland, such blocks look quite out of place and the farmer should try to improve the design, in several respects, in future planting.

In a landscape of fields bounded by fences and hedges, the location and shape of true shelterbelts is virtually dictated by the present field boundaries and they will be set at right angles to the most damaging wind direction in order to be of most use. In sharp contrast, on upland and hill farms where the landscape is dominated by landform rather than by field fences, the location and shape of woodland blocks can be selected more freely. Thought

should be given to ensure the planting is fitted to the scale of landscape and is an asset – for farming, for conservation and 'for the look of the place'.

Try to avoid the simple rectangular block, especially on the face of a hill; the changes to achieve a more natural shape need only be quite small and will be well worth the extra fencing cost.

The location of tree blocks on a grazing farm is not something to be left to a landscape architect working in a studio because they must take account of the way the animals, especially sheep, move on the ground, particularly in bad weather; a badly sited fence or block of trees may trap sheep and cause severe losses. Provided that such factors are taken fully into account, it is helpful if the planting design is linked to some land feature or features, perhaps a hollow in the hill, and if the outline follows the shape of a local landform.

When tree blocks are being planted in arable land, consideration should be given at the planning stage to access for transport, not least to the removal of the timber when thinning is due. It may be that this is planned to be across an arable field in the interval between cereal harvest and autumn ploughing. That may be adequate for a very small wood, but remember that a ten-hectare wood on fertile soil, as it will be on a lowland arable farm, could well produce a thinning yield in excess of 500 tonnes of wood on each visit, which could not be taken out in a few days. Make provision for proper access to woodland, even if the production is not on this scale.

If planting is to be done on land with tile drains, remember that the tree roots will block them. It may be this will affect the drainage on neighbouring fields. On a new planting site any necessary drainage should be by open ditches.

TIMBER
PRODUCTIVE
STAND

OPEN-GROWN
MARGIN
best if native
broadleaves

useful for wildlife and
as space for stacking
and manoeuvering

SOFT EDGE OF A TIMBER BLOCK

It is unwise to attempt the ultra-clever design. The rich, newly arrived foreign sportsman-farmer who planted his initials a quarter-mile high on a grassy hillside in the Scottish Borders and, for agriculture-grant purposes, called the results 'shelterbelts', will be long remembered – with disdain.

The species of trees to plant in shelter blocks deserves careful thought. On a hill farm, if the planting is small and has no road access, it is possible it will never be harvested. There is then no commercial cropping influence on the species chosen, simply consideration of what will perform best. The grants offered by the Forestry Commission are currently weighted strongly in favour of broadleaved trees, thus providing help towards the planting of native trees (including Scots pine in the Caledonian Forest area) which are particularly well suited to hill plantings for shelter.

On exposed sites the outer rows of trees inevitably become stressed, often stunted or deformed. Even when the farmer designs the planting of larger blocks with a view to a commercial harvest, it is good practice to include a range of native broadleaved species in this outer fringe in order to increase the diversity, the wildlife interest and the stability of the main crop, with no loss of harvestable volume and probably with an enhanced grant because of the increased broadleaved element.

Trees on the farm are not only for timber

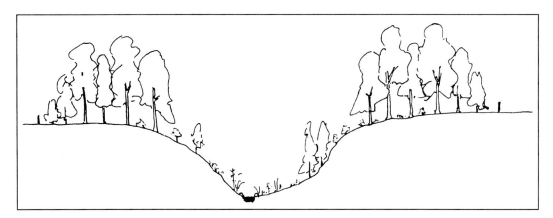

PLANTING A GULLY
Keep planting back from the streamside to let sunlight reach the water for fish and other wildlife. Timber trees on steep slopes may be difficult to harvest anyway.

production and shelter. They can be planted with great advantage in the rounded-off corners of fields, where big farm machines either cannot get into at all or spend so much time reversing and manoeuvring that taking the corners out of cropping makes a clear profit. These corners can be planted with a few trees to form part of a focal point for wildlife, part of a system of linear habitats across the whole farm. This has been well demonstrated on some of the country's most successful arable farms, now swinging back from the practices which had previously stripped them of every tree and most of their hedges.

Although there may be farming advantage in fencing off and planting deep gullies on upland farms – cleughs in Scots, ceunant in Welsh – where sheep are prone to become buried in snow and cattle may fall, care should be taken of the streamsides which usually have special conservation value in such places. This is a great opportunity to improve wildlife habitats, at no cost to farm production, with planting of native trees and bushes which were the normal vegetation there before intensive grazing changed them. The species to plant will vary across the country and with altitude; birches, rowan, ash, aspen, sessile oak, holly and common alder will often feature. In general, the tree-planting should be kept back from the streamside, leaving space for tall herbs which will probably arrive by self-sowing; leaving the area open should not reduce the planting grant from the Forestry Authority, since there is provision for this in the scheme. The shelter and tall-herb fringes are especially important in supplying food for fish and other water life. Dense shade by trees over-arching the water, especially conifers, is ecologically unhelpful and should be avoided.

AGRO-FORESTRY AND SILVI-PASTORALISM

There has been some renewed interest in land-use systems involving close mixtures of traditional farm crops and trees, but in Britain they are not simple ways of getting rich quickly.

More than 50 years ago the match-makers Bryant and May successfully raised plantations of poplars by setting out the plants at wide spacing in fields of wheat; only when the trees were well established and beginning to shade the cereal did crop yields fall appreciably. This silvi-arable system went into disuse with the import of poplar logs and then of cheap matches, but in its day it was successful and it is now being practised again, with narrow-crowned hybrid poplars planted at 6m x 6m or 8m x 8m over either food crops or kale for pheasants.

In the sub-tropics and tropics, where the sunlight is more intense than in Britain and, critically, more intense than many important food crop plants can use, agro-forestry systems are common: soya beans grown under poplar, cocoa and coffee grown under high-shade trees, oyster mushrooms grown under pine plantations, and so on. The intensity of the solar radiation, both light and heat, is the critical factor, and there are few situations in the north of the north temperate zone where farm crops benefit from the permanent shade of trees. The cultivation of chanterelle fungi or oyster mushrooms might be examples, if someone wants to develop an unusual alternative crop.

Silvi-pastoralism is the grazing form of agro-forestry and a simple extension of woodland pasture systems and parklands which have existed since time immemorial. Special point was given to this development by the success of trials in New Zealand leading to its widespread adoption there. Using the radiata pine (*Pinus radiata*), New

Zealand farmers have been able to raise excellent timber crops on land which has also provided good grazing for at least two-thirds of the tree crop rotation. In effect, from the time of planting the trees, the grass production declines from 95 per cent of the maximum possible in an open field to 20 per cent at year 20, when the tree harvest is the same as that obtainable on the land without any grazing. The land really does provide two crops simultaneously.

Here again, the critical factors are the solar radiation, the basic climate and the growth rate of the trees. With a long growing season (300 days a year is usual over much of the country), radiata pine can be grown in New Zealand to large sawlog size in 20 years or so, a much shorter production period for timber than is possible in Britain. After the trees have been planted at wide spacing between the rows, the grass is cut with a forage harvester for two years, grazed with lambs for a further year and then opened to adult stock. The trees are high-pruned for a top-value sawlog product. Fast-growing radiata pine is safe from sheep-browsing and rubbing by year three, much sooner than is the case with pines in Britain, where protection for individual trees is essential for several years, which is costly. Nevertheless, there is sufficient promise and possibility in the system to justify research, which is in hand, especially in Wales and Scotland. High-value broadleaved trees appear to offer the best promise: gean, sycamore and hybrid poplars, although the conifers, hybrid larch and Douglas fir may also be useful.

In view of the slower growth of trees in Britain than in New Zealand, it is necessary to protect trees from animal-browsing for a longer period. This may be done with treeshelters, if the grazing animals are sheep. Tall shelters are needed, 1.8m, and they must be double-staked to withstand animals rubbing on them.

The husbandry must be designed with two principal aims: to ensure that the trees provide produce of the highest possible unit value, and to ensure that the trees interfere as little as possible with the pasture. With these objects in mind, the trees should be tended to grow high-quality wood as quickly as possible; mulching and high pruning are needed. Pruning not only improves the potential value of the timber but it allows more light to reach the grass than would be the case with low branching. Experience shows that the shelter provided by the trees generally allows the grass to begin growing slightly earlier in the spring and to grow later in the autumn, thus in part offsetting the reduction of the grazing area.

Other than with hybrid poplars on very fertile land, silvi-pastoralism is experimental in Britain; it may prove attractive but it is too early to recommend it.

Many farmers do make use of their woodlands to hold stock in shelter during severe winter weather, although this usually requires supplementary feeding, since woodland grasses are notably less nutritious than open sward species. Two points of importance arise. Woodland which is heavily grazed for long periods (i.e. more than about eight sheep and deer per 100ha) quickly becomes denuded of its most palatable plants and of tree seedlings; hence it is made unsustainable and, if the grazing continues, will gradually degenerate and disappear. This has been the major cause of the decline of important oakwoods in the Lake District, in the west of Scotland and in Wales; in all these places there is now a programme of protection promoted by the nature conservation agencies aimed at ensuring regeneration. Continued use of the woods by sheep and cattle in winter should be possible so long as it is with moderate stock numbers and provided that there is complete protection for several years periodically to allow tree seedlings to be established and to grow beyond the size when they can be eaten.

The other way in which farm woods are used by stock is to provide shelter in which animals may be held for convenient hand-feeding, without any reliance on woodland vegetation. Sometimes a small area is fenced off, a hectare or less, and a heavy concentration of animals is held. The result is usually the destruction of the woodland, the ground being compacted and tree roots exposed. If the system is to be adopted, the farmer should make clear arrangements for the rotation of the enclosure, so that the trees have several years to recover from the harsh treatment, not only from trampling and gnawing of bark but from the extraordinary (perhaps lethal) amount of manuring. Discretionary grants are now available, currently up to £80 per ha, as Livestock Exclusion Annual Payments, for keeping animals out of 'conservation woodlands' in less-favoured areas or environmentally sensitive areas.

It is not only cattle and sheep which may cause such problems. At the estate of the Rural Economy Centre near Edinburgh some years ago, half a hectare of mature trees were killed when it was decided to pen 10,000 hens temporarily in woodland adjacent to a poultry research station. The trees were pecked to death and poisoned with droppings.

A great impetus has been given to new woodlands on farms, including short-rotation coppice, by the EU regulation which allowed farmers, from July 1995, to count arable land in forestry schemes towards their set-aside obligations under the Arable Area Payments Scheme (AAPS). To qualify, such land must not have been in permanent pasture, perennial crops, woodland or used for non-agricultural purposes. Under the new regulations, set-aside land is eligible for full Woodland Grant Scheme (WGS) payments from the Forestry Commission, subject to the usual conditions, and annual payments of the Farm Woodland Premium Scheme (FWPS), which are up to £300 per ha for 15 years. The agricultural departments and the Forestry Authority administer the grants schemes. The land has to meet the normal AAPS set-aside rules: that applicants will normally have farmed the land for at least two years, and the land must have been cultivated with a view to harvest or have been in set-aside (including the old Five-Year Set-Aside Scheme) in the year before application for tree-planting. The minimum plot sizes are not likely to be a constraint. The former requirements that set-aside land should be kept in good arable condition which prevented its lucrative use do not apply, and the accepted land which is planted with trees can be counted towards a farm's obligation under the flexible set-aside requirement. These changes provide a huge encouragement to effective farm-forestry in Britain.

In recent years a quiet revolution has been taking place across the British Isles, so that many farmers are perceiving values – not only money – in their woodlands and are becoming active in their renovation. Encouraged by new woodland grants and by the emergence of increasingly effective local woodland associations, farmers have begun to put trees and multi-purpose woodlands back on the farms. The 'art is long', so there is no sudden change in the landscape, and there will be continual cause to strike the middle road between the creation of pulp slums and dilettante contrivings which otherwise would be the derelict woods in decades to come.

In this respect, the Woodland Improvement Grants paid by the Forestry Authority are particularly helpful for farm woodlands, encouraging the renewal of those which have fallen into disrepair. Woodland Improvement Grant Project 2 is a discretionary grant aimed at supporting owners wishing to bring woodlands which are of low commercial value and in poor condition into management, especially small, isolated broadleaved and mixed woods on farms. Typically it assists operations such as selective felling, coppicing and conversion to high forest, rhododendron control and protection which should promote sustainable management.

Woodland Improvement Grant Project 3 is aimed at increasing biodiversity in woodlands to promote habitat conservation for wildlife, especially in semi-natural woods. The preparation of management plans, coppice management and the positive conservation of native woodlands are among the works eligible for funding (see appendix 1).

A tide of renewal in farm woodlands is beginning to flow; may it further increase.

FURTHER READING

Blyth, J., Evans, J., Mutch, W.E.S. & Sidwell, C. (1991), *Farm Woodland Management*, Farming Press, Ipswich

Forestry Commission (1991), *Edge Management in Woodlands*, Occasional Paper 28, Forestry Commission, HMSO, London

Forestry Commission (1992), *Forests and Water Guidelines*, HMSO, Edinburgh

Forestry Commission (1992), *Lowland Landscape Design Guidelines*, Edinburgh

Introducing Farm Woodlands (1998), Published jointly by the Forestry Commission and the Agricultural Departments (MAFF, SOAEFD and WOAD); available free from Forestry Authority Conservancy Offices

Mutch, W.E.S. & Hutchison, A.R. (1980), *The Interaction of Forestry and Farming*, Economics and Management Series 2, East of Scotland College of Agriculture, Edinburgh

Hedges

13

Although the removal of hedges nationally has often been considered in terms of their conservation effects, it is well to remember that hedges serve several functions apart from providing wildlife habitat. The objectives influence their composition and management. Whether clipped or laid, hedges are the most artificial woodland-type habitat, being entirely contrived by man.

The objectives of hedge management may be:

- to confine farm stock
- to provide shelter from wind for stock, field crops or orchards
- to provide special linear habitat for wildlife, principally birds and insects, but also plants
- to filter weed seeds and blowing soil from the air
- to trap snow

To serve these roles, some special qualities are required for a tree or shrub if it is to perform well in a hedge:

- It should be stock-resistant, therefore tough (not brittle or delicate).
- It should be unpalatable to stock, so spines and thorns are an advantage, but it must not be poisonous.
- It must stand clipping and be a compact grower.
- It is helpful if it coppices well or layers, as aids to regeneration.
- It should not spread into fields by aggressive rooting.
- It should be a robust plant, not frost-tender or prone to fungal attack.

Three common species are ruled out by being poisonous: yew and box (poisonous foliage) and laburnum (poisonous seeds). Others which are generally too invasive are rhododendron, broom, elder and whin (gorse); the last-named does serve as an informal hedge in some parts of the country, but its rampant aggression and persistence of seed in the soil make it difficult to recommend.

Conifers do not coppice or layer but a few are sometimes used quite effectively as hedge plants; Lawson cypress and western red cedar are the best, forming tight, dense hedges if kept close-clipped. They are suitable for some gardens and nurseries. Scots pine hedges exist in East Anglia and occasionally elsewhere and spruce occur in some west-coast areas, but neither can be generally recommended. Leyland cypress has been planted very widely in recent years, especially in suburbia, but is a recipe for disaster and should be avoided for hedging; its easy establishment and rapid early growth make it easy to sell but those who are fighting desperately to hack back a hedge already 7m tall, 4m thick and with aggressive roots spreading 5–6m across their garden do not need to be told that it is unwise to plant a tall-stature forest tree as a hedge. Nor does Leyland cypress have any conservation merits.

For general use as a clipped hedge or a laid hedge, the prime species are:

Beech	Hawthorn
Hornbeam	Midland thorn
Holly	

For hedges to be managed by laying (see below), some of the following species may be added as occasional constituents:

Field maple	Cherry plum
Dogwood	Damson (bullace)
Spindle tree	Blackthorn
Hazel	Dog rose
Wild apple	Guelder rose
Crab apple	Ash

On sandy soils and in salty conditions (seaside and beside traffic roads), the sea buckthorn and Ramanas rose are essential.

On chalk and other calcareous soils, the wayfaring tree and common buckthorn are likely additions, with beech, hawthorn and holly the major species.

On heavy clay soils, hawthorn, holly and blackthorn will be the principal species, with damson (bullace) and gean as occasional additions.

Willows are useful for the creation of hedges required to give quick results, especially on moist soils. They should be backed by another hedge of long-lasting species. In wet west-coast areas with severe exposure, the same purpose of forming a quick and probably temporary hedge may be performed by Sitka spruce, but if cutting is neglected it later grows into a very ugly line of scruffy trees.

NEW HEDGES

Having decided on the exact line of the new hedge, choose the species to be planted, taking full account of what succeeds locally. It is necessary also to consider how the hedge is to be managed, whether by trimming or by laying. Laying involves allowing the hedge to grow upright for some years, then part-cutting through most of the shoots close to the ground, trimming off the side branches from them and weaving them almost horizontally along the hedge through occasional upright stems and stakes, to create a plaited living hurdle a metre or more high. In deciding the line of the hedge, consider the sight-lines for road traffic and the visibility of people and vehicles coming through a gateway on to a road.

Plants may be home-grown but are readily obtained from reputable nurserymen. Care should be taken to get plants of native (local) origin; particular care should be taken to obtain native hawthorn, not some Central European provenance supplied for the nurseryman's convenience.

If possible, the line of the hedge should be ploughed with a sub-soiling tine to 60cm depth. The vegetation to a width of 1.5m on the hedge-line should be killed with a pre-planting herbicide; glyphosate (Roundup) is satisfactory, or use a plant desiccant such as paraquat/diquat. Herbicide treatment is especially necessary if the vegetation is a grass sward, which can virtually prevent the plants growing at all.

The hedge, once grown, will be at least 1.5m thick, 2.5m or more for a double-row conservation hedge; allow space for this.

For a stock-proof hedge, the plants may be set in a single row, with 25cm between plants or, better, in a double row, with 30cm between plants and 40cm between the rows. The plants will probably be at least 60 per cent hawthorn in most districts.

For a hedge largely intended for wildlife, there should be two rows, 45cm between plants and 75cm to 1m between the rows. The planting might be 50 per cent hawthorn and 10 per cent each of five other suitable shrub species, the choice depending on the locality.

Plants should be well-balanced, with about

40cm shoots and good fibrous roots. Cell-grown plants are very suitable, although rather more expensive than bare-rooted. Winter or early-spring planting is generally best in most places. Hawthorn will stand the shoots being 'topped' (pruned back to 40cm height); use stocky plants with thick root collars, not thin wands.

After planting, the hedge must be kept weed-free until the plants are growing vigorously. This may best be done with a herbicide such as simazine or propyzamide (Kerb). Pay close attention to the rules for using these chemicals. Whatever method is used, the hedge plants must be kept clean-weeded in order to encourage vigorous growth.

HEDGE PROTECTION

If there is stock in the fields adjoining the new hedge, there is no alternative to fencing the line (perhaps on both sides) to keep animals off the growing plants for four years or so. An electric wire fence or electric sheep netting fence may be sufficient; whatever fence is used, it must be set back so that animals cannot reach across and get at the hedge.

In areas without farm stock, fencing may not be necessary. However, hares and roe deer will eat plants; immediately after planting, hares may go down the line and bite off every plant in turn, laying the severed shoot neatly beside the stump. Nevertheless damage by hares and roe deer is unlikely to be disastrous and can usually be held at an acceptable level by local shooting. It is also particularly easy to put sheep's wool on a young hawthorn hedge (see chapter eight) and this may be sufficient deterrent for roe deer. Very serious damage may be done by rabbits and, in grassy swards, by voles. Rabbit fencing is expensive; rabbits should be rigorously controlled by shooting and poisoning. They may be deterred by mulching the new planting, either planting through a strip of black plastic sheet or applying straw or wood-chips. Although the

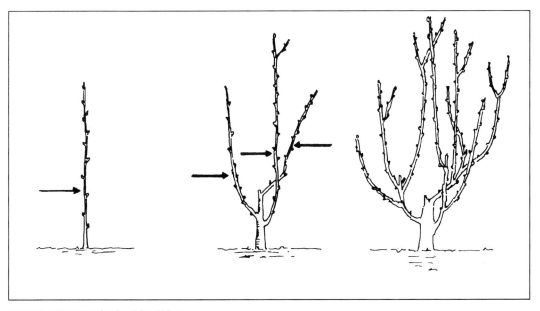

EARLY PRUNING IS ESSENTIAL
Young hedges must be forced to branch repeatedly low down.

use of quills is sometimes suggested (see chapter eight) as a protection against vole damage and moderate protection against rabbits, they do not prevent a mature rabbit biting off the plant above the top of the quill and their use prevents the timely pruning back of the hedge plants to induce low branching. Vole damage is greatly reduced, and may be quite avoided, by pre-planting herbicide treatment of the hedge-line and by subsequent clean-weeding.

When the hedge is established it should be allowed to grow but pruned back in the first winter to 60cm to encourage low branching. It may then grow on and be cut back to 1m. The final height for a trimmed hedge may be between 1.4m and 1.8m.

Trimming should be done in late winter, preferably every second year. The most suitable shape is usually a flat-topped A, which can be maintained with a tractor-mounted cutter or by hand.

Trimmed hedges, even when carefully managed, tend to become gappy and 'leggy', and eventually may consist of isolated bushes with tight trimmed heads on tall stems, not stock-proof and providing inadequate shelter for wildlife or crops. When the hedge shows signs of so developing, the best practice is probably to erect a temporary fence or fences and to cut back the whole hedge nearly to ground level (i.e. to coppice it), filling in any gaps with new planting.

Hedges maintained by laying are very neat and slim when newly laid but much bulkier than a trimmed hedge after some years; they are more reliably stock-proof (and person-proof!) than a trimmed hedge. Laying may be necessary on a rotation of about 15 years. At the time of laying the main stems should be 50–100mm thick at the base, and up to 3.5m tall, having been allowed to grow to height for two or three years. The process, which should be done in winter, is a skilled one. It begins by

FLAT-TOPPED A-SHAPE
If space is not at a premium, the sides may be cut at a gentler slope, which improves the wildlife value. Even if there is no ridge on the ground when the hedge is planted, one often forms due to blown soil and debris trapped by the hedge's shelter.

cleaning out useless material such as brambles, elder and deadwood, and trimming drooping side branches. Then long stems at regular close spacing are selected as 'pleachers', which must be cut half through near the base to allow them to be bent almost horizontal (i.e. 'laid'). Other stems are cut off at about 1.5m height to act as uprights (supplemented by cut stakes) and the pleachers are then plaited into the uprights. If the ground slopes, the pleachers are laid uphill in order to reduce the bending of the half-cut stems. A skilled hedger produces a neat result like a well-made basket. After laying, no trimming should be done for two years, to allow the wounds of the pleachers to heal and the new hedge to establish.

HEDGEROW TREES

In many parts of the country it is traditional to grow timber trees in the line of the hedges,

LAYING
Hedge laying is relatively straightforward where there is an abundance of straight shoots; it is difficult where there are a few gnarled stems. The uprights may be retained growing hedge shoots which will sprout again from the top, or driven stakes.

so that they become 'hedgerows'; these trees are highly desirable not only for their landscape and wildlife habitat benefits, but also as timber resources. The species involved are almost always broadleaves. Trees such as beech, which cast particularly heavy shade, tend to weaken the hedge beneath their crowns. Ash, oak, gean and sycamore are species well adapted to growth in hedgerows, the choice depending on the part of the country. Where the hedge becomes weak in the shade of the trees it may be thickened using a shade-tolerant hedging species, holly or blackthorn, instead of hawthorn. If the hedge is to be trimmed with a tractor-mounted cutter, each young tree should be clearly identified with a bright fertiliser bag tied to it, or some such marker, to alert the driver.

Hedgerow trees will probably have to be pruned, both for the production of a good butt log and to raise the crown in the interests of farm crops and grass. Some loss of crop production and grazing has to be considered. It is seldom appropriate to have hedgerow trees at closer spacing than 30m. In some hedges maintained by laying it is usual to recruit at least some of the trees by allowing a shoot, especially ash, to grow on instead of becoming a pleacher, analogous to storing coppice.

SNOW AND HEDGES

Hedges have a marked effect on the drifting of snow; in windy conditions, when snow drifts, it is not unusual for the fields to be completely clear and the roads, between hedges, to be totally blocked with two metres of snow.

A semi-permeable screen causes the snow to be evenly spread over the leeward field (the principal aim of shelterbelt planting in the Great Plains of the USA and the Russian steppes, in order to provide water for the cereal crops in the following year). In contrast, a dense barrier causes snow to accumulate in the narrow zone of deep shelter close to the hedge itself. This should be carefully considered when planning a new hedge or shelterbelt; it may be best in snowy districts to plant 30m or 40m back from the farm access road (as roads authorities build snow-fences), so that snow drifts form where they are not an inconvenience and access is left open.

FURTHER READING

Hodge, S.J. (1989), *The Establishment of Trees in Hedgerows*, Research Information Note 195, Forestry Commission, Edinburgh

Maclean, M. (1992), *New Hedges for the Countryside*, Farming Press, Ipswich

Pollard, E., Hooper, M.D. and Moore, N.W. (1974), *Hedges*, William Collins & Sons, London

Special Products – Special Crops 14

SHORT-ROTATION COPPICE

A new crop, just appearing in Britain and other countries of northern and western Europe, may be of special interest to farmers: short-rotation willow or poplar coppice. The interest arises from the coincidence of three concerns: the over-production of food in the European Union leading to farm set-aside; the national need for truly sustainable sources of energy; and the threat of global warming associated with the increase of carbon dioxide in the atmosphere. The crop is half-way between forestry and arable farming.

Since the price of oil and natural gas is expected to rise substantially in the long term, and nuclear power is suspect, justly or unjustly, on environmental grounds and economically, owing to the costs of decommissioning nuclear power stations, the search for renewable sources of power has been given urgency. The UK government created the Energy Technology Support Unit (ETSU) in 1979 to investigate biomass fuels which might be burned in conventional stoves, supply district heating and power schemes, or generate electricity with an engine burning gasified wood.

The first of these objectives may serve people heating private houses, farms, etc. The second depends on decisions of local authorities, hospital boards and industrial firms to participate in district heating. The third depends on national bodies and, specifically, requires investment in an engine burning gasified wood.

In considering species for biomass crops, plant breeders in Scandinavia have concentrated for many years on the willows, while those in Belgium and Italy have concentrated on poplars. Sweden has much experience of district heating based on wood burning and aims to create 200,000ha of coppice willow by the year 2000 (in addition to massive existing forest residue resources), enabling the country to dispense with nuclear power generation.

A large collection of willow species and hybrids was planted at the research station at Long Ashton in Somerset, including those from Sweden, and many trial plots were planted there. The most promising for Britain are selected or bred from *Salix viminalis*, the common osier.

For this crop, willow should be managed as coppice on a three-year rotation; poplar should also be on a three-year rotation, or perhaps even on two years on the most productive sites. The crop is established initially using unrooted cuttings 20–25cm long and 8–16mm basal diameter, ensuring that each carries several strong vegetative buds and not exclusively flower buds. Several willow cultivars may be planted together on a site to provide biodiversity as an insurance against disease. The most suitable poplars for short-rotation coppice are probably *Populus trichocarpa*, 'Fritzi Pauley', and *P. trichocarpa* x *P. balsamifera*, 'Balsam Spire'. Cuttings should be set in the ground with about 25mm showing above the soil, and 75cm between cuttings. The spacing between the rows is alternately 75cm and 150cm, i.e. the rows are in pairs 75cm apart, with 150cm between each pair (12,000 plants per ha). Tractor access, straddling the rows, is required for initial weed control and for harvesting; the planting system of 75cm/150cm, now standard in Sweden, Germany, Ireland and the UK, is important in ensuring the crop matches the industry standards for machine-harvesting.

Weed control is essential. The grower must ensure the crop suffers no competition and closes canopy as soon as possible. The site should be given pre-planting herbicide treatment and cultivated to 40cm depth. It must be kept weed-free in the first and second years. The growing cuttings should be cut back at the end of the first year in order to encourage basal growth as coppice. In the first year at least the crop requires protection from roe deer.

Although traditional coppice is not given fertiliser, most three-year coppice systems will require either superphosphate, phosphate/nitrogen or balanced NPK fertiliser, depending on soil analysis. Short-rotation coppice is an excellent crop for the application of slurry, which can substitute for mineral fertilisers. It is environmentally safer to apply slurry to coppice than to stubble.

Harvesting on a field-scale is done by mechanical coppice harvesters, of which several are now available. The details of the preparation, bundling, required chip size, etc. are fixed by the end-use customer. It is important that the coppice stools are left in good condition for sprouting to provide the next crop. Cutting may be done at any time in the leafless period, say from November to March. Co-operative purchase and sharing of a harvester appears to be feasible and prudent; the precise date for harvesting is not critical within the winter, so a machine can serve several growers without any inconvenience.

In order to achieve economically useful yields, short-rotation coppice management should be attempted only on fertile soils of arable cropping quality and in a moderate to mild climate; it is not yet clear how far north the crop may be grown successfully. Yields on suitable ground should be in the range ten to 20 tonnes of dry wood per annum (i.e. a harvest of 30 to 60 tonnes of dry wood at three years; it is prudent to be conservative in the planning stages).

Clearly, before a crop is planted for sale, careful market research is required to establish the potential customers' specifications and the prices to be paid, as well as the growth prospects. For use as domestic fuel, the following figures may be helpful. A three-bedroomed house consumes about 130 giga-joules (Gj) of energy per annum, the equivalent of some 3,000 litres of heating oil (660 gallons) or 4.3 tonnes of coal. This could be supplied by nine tonnes of air-dry wood (25 per cent moisture content – mc) or seven tonnes oven-dry. Drying the chips to 20 per cent mc or less is most important; freshly cut wood, at say 60 per cent mc, has a heating value of only 6.3 Gj per tonne (less than half that of the dry), since much energy must be used in evaporating the water; in addition to that deficiency, burning the wood green results in abundant tar condensing in the cool upper chimney, which may then catch fire. Space must be made for drying the chips and storing them; allow 2.5m³ per tonne. Scandinavian farmers run chippers from the power-take-offs of farm tractors for home-heating chips. With a yield of only 9 tonnes (air-dry) per ha per annum run on a three-year cycle, coppice might supply a three-bedroomed house with its energy from a plantation of 1ha, one-third being cut each year.

It must be stressed, however, that this is a

crop for fertile, arable soils, perhaps in set-aside, and the suggested yields have been proved only in southern England and in the mild west. At present the willows bred in Sweden are likely to succeed better in northern England and south Scotland than the Belgian poplars. It may be that new northern strains of poplar will be required for the north. The Forestry Commission research division has a network of 48 trial plantings across the whole UK, so farmers should soon have hard information on feasibility and suitable strains.

The crop is environmentally friendly. In appearance it resembles maize. Since it is grown in arable fields which are already set out as rectangles, it is not visually unusual and its colour and texture may enhance diversity. It requires much less fertiliser than most agricultural crops and, normally, no fungicide or insecticide, nor, after initial establishment, any herbicide. The fact that each coupe is an undisturbed sanctuary for three years is ecologically helpful.

It seems that much more will be heard of this crop in future. The decision of an EU Council in 1995 that farmers may count arable land entered into forestry schemes towards their set-aside obligations under the Arable Area Payments Scheme gives great impetus to this type of crop.

HIGH-PRODUCTION POPLARS

Strictly, it may be held that poplar wood is not a special product but nowadays the crop is such a specialised one that it warrants a place here, particularly for the interest and guidance of farmers who have not previously considered it.

There are about 30 species in the genus *Populus*, all belonging to the north temperate and sub-arctic zones where most grow on moist soils on river flood plains. They hybridise readily. Some are very prone to a bacterial infection (*Xanthomonas populi*) which causes severe cankers on the stem, and to fungal diseases of the leaves, including rusts (*Melampsora spp.*) which weaken the tree and slow its growth; other poplar species are resistant to these diseases, so selecting and breeding disease-free strains pays high dividends in poplar research.

Poplar wood is white to pale yellow-brown, lightweight (400–550kg per m^3), normally straight-grained, soft and easy to slice and carve. The sapwood and heartwood are not well defined. The wood is without smell or taint, being useful therefore for carrying food. It saws and finishes quite well and is about as strong as spruce, so it might be thought acceptable as a structural timber, but it is prone to twist and warp when seasoning and is generally not used for structural work. For many years poplar has been veneered and used for fruit baskets and there is a demand for vegetable crates. Logs are veneered when green, i.e. unseasoned, and before shrinkage of the wood occurs.

Now there is increasing demand for fast-grown logs to saw, to pulp and to chip for panel products, as well as interest in the possibilities for growing special cultivars on a two-year or three-year cycle for wood fuel chips, as described in the preceding section. Poplar's fast growth is the attraction for all these markets.

Internationally there is intense interest among farmers and forest services to find highly productive strains of poplar which are resistant to bacterial canker and rust diseases. Much of the research is done in Italy and Belgium. Because of the threat of the diseases, it is a legal requirement under the Forest Reproductive Materials Regulations 1977 that plants and cuttings of poplar sold for wood production should be obtained from a registered source, for which the

Forestry Commission is the UK authority.

The acceptable clones registered in 1997 for use in the UK fall into five groups. The first is *P. x canescens*, the grey poplar, which is a natural hybrid between the white poplar (*P. alba*) and aspen (*P. tremula*). It is frequently planted for amenity, having attractive foliage, and tolerates a wide range of soils, provided they remain moist through the summer. It grows well on fertile soils even in north Scotland and produces good timber.

The remaining four groups are various hybrid combinations of:

Populus nigra, the European black poplar, which is highly resistant to the bacterial canker;

P. deltoides, the eastern cottonwood of North America;

P. trichocarpa, the western balsam poplar of North America, with a wide range from Alaska to California; and

P. balsamifera, the true balsam poplar of North America (previously having the unusual Indian name of *P. tacamahaca*).

The hybrids of *P. deltoides x P. nigra* are properly called *P. x canadensis* but usually and more descriptively *P. x euroamericana*. Hybrids of *P. trichocarpa x P. deltoides* may be called *P. x generosa* but more usually *P. x interamericana*.

The registered named clones of *P. x euroamericana* are 'Robusta', 'Serotina', 'Eugenei', 'Gelrica', 'Heidemij', 'Casale 78', 'Primo', 'Ghoy', 'Gaver' and 'Gibecq', but only the first and the last three are now planted in any quantity in Britain. They are impressively productive on the right sites and, being small-leaved, stand strong winds better than some other strains.

The registered clones of *P. x interamericana* are 'Beaupré' and 'Boelare', which are very fast-growing and of excellent form.

There are four registered clones of pure *P. trichocarpa*: 'Fritzi Pauley', 'Scott Pauley', 'Trichobel' and 'Columbia River'. They are not so fast-growing as the previous hybrids and may pose problems with stem breakage in high winds, but they have very fine form and are useful in high-rainfall areas. 'Fritzi Pauley' is probably the best and the most widely used; unfortunately it is rather prone to growing epicormic shoots after pruning and has rather large leaves for windy areas.

The single registered clone of the *P. trichocarpa x P. balsamifera* cross is 'Balsam Spire', widely used because of its elegant narrow crown and its habit of early flushing, which is useful for sheltering farm crops.

Commercial planting of poplar must be confined to free-draining base-rich loam soils in sheltered locations usually below 100m elevation, with a water table 1m to 1.5m down in summer. Waterlogged soils should be avoided, as should those which are more acid than pH 5.0 and shallow soils (50cm or less). Arable land in permanent set-aside under the EU's Common Agricultural Policy may be suitable. Arable fields which may have a plough pan should be ripped with a sub-soiler. Upland sites and exposed sites are unsuitable. Trials have not yet shown whether these high productive crosses can be grown in the north of England and Scotland; at present they cannot be recommended there.

It is possible to plant rooted cuttings but these are expensive to buy and require costly pit-planting. It is more usual practice to use unrooted cuttings. For wide-spaced planting (8m x 8m), use 1–2m cuttings of two-year-old straight shoots, at least one-third below the soil surface. For higher density planting, shorter cuttings, even 25cm, may be used to reduce costs. Cuttings must not be thrust into the soil (the bark and cambium will be damaged and rooting inhibited) but put into a prepared hole; a tractor-mounted soil auger and a crowbar may be used, but care must be taken that the cutting is not left dangling in air space with only the top of the hole filled in. Set cuttings in early spring.

Past practice has been to set cuttings at very wide spacing (8m x 8m, only 156 per ha), which requires every plant to grow and allows no thinning, for felling at about 25 years. Increasingly, growers now plant closer, giving the advantages of early thinning and a choice of trees for the final crop. Planting at 2m in lines 4m apart gives 1,250 cuttings per ha. Land should be tine ploughed to break any plough pan and either herbicide-treated or cropped inter-row with beans or cereals (maize or soya are normal abroad). Soon after the cuttings have flushed, buds on the lower stem may be removed by hand as a first pruning. In the autumn and following spring, cultivate inter-row to reduce weeds.

For the sawlog and veneer markets, poplars must be pruned annually (or at most biennially) from about year three, to a height of at least 6m. (If pruned too severely at one time several clones are apt to produce heavy growths of epicormic shoots, which ruin the trees.) Pruning is best done in late summer. Thinnings are usually in years six, eight, ten and 15, reducing the stand to about 180 stems per ha for final felling at about year 25 (may be year 30 on poorer sites), with the crop trees about 60cm diameter at breast height (dbh; see page 254) and a final yield approximately 550m³ per ha.

Populus trichocarpa (mostly 'Fritzi Pauley') and the *P. trichocarpa x P. balsamifera* 'Balsam Spire' are being used at closer spacings, down to 2m x 2m, to produce high yields of pulp and fibre on rotations of about ten years. For this they are not pruned or thinned but are cut as poles (average 25cm dbh) in the normal way.

Provided that at least 1,100 trees per ha are planted, full planting grants are paid for the approved poplar clones under the Woodland Grant Scheme and Farm Woodland Premium Scheme (see page 304). If fewer trees are planted, grants are paid pro rata.

Before deciding on poplar growing as an enterprise, farmers should take expert advice on the suitability of the site and on the most promising hybrid and clone, and should consider the markets. Veneer logs will stand long-distance transport to mills but fuel chips require a much closer market.

WALNUT

Until mahogany logs were imported into Britain in large quantities in about the middle of the eighteenth century, the premier wood for high-quality furniture and carving was walnut. With the rapid destruction of the natural forests of mahogany and other decorative hardwood trees in the tropics, the furniture industries of Europe in future will have to rely increasingly upon home-grown timbers. Consequently, there is a high probability that the demand for walnut will strongly increase, as will its price, which is already firm.

The second factor encouraging more attention to growing walnut is the trend in climate. Whether the gradual warming of our land is attributed to a long natural cyclic movement or to a Man-induced greenhouse effect, it is helpful to the culture in Britain of this tree, which is native to the Balkans eastward.

With the prospect of rising price for the timber and the slow climatic improvement, however, walnut culture remains a speculative venture and one which must be confined to the best soils and favourable situations. This can be a peak of quality timber growing, which should be the watchword for small woodlands.

The soil for good growth of walnuts must be a fertile, deep, moist loam. Heavy clays, waterlogged soils, poor sands and shallow soils must be avoided. Although walnut is winter-hardy, it suffers in late-spring frosts and, for this reason, it should not be planted in low-lying frost pockets. Areas of heavy wind exposure should be avoided. Local shelter and deep fertile soil are more important than latitude; excellent trees of large size grow as far north as the Moray Firth.

Plants can be grown from nuts sown as soon as they are ripe in early autumn, or the nuts may be stored over winter, stratified in damp sand outdoors. If possible seed should be taken only from timber-shaped trees rather than from wide-spread specimens which may have been selected for their nut-bearing. If nuts are to be stored by being stratified (i.e. 'sandwiches' of nuts and sand alternately, layers of nuts two-deep and two inches of sand), they must be fully protected from mice and other vermin; a strong wooden box with bottom and top of strong mouse-proof wire net will do well, exposed to all weather. Walnut plants develop a strong tap-root and, for that reason, are difficult to transplant without a severe setback. Indeed, old forestry wisdom is that transplanted trees which have lost their tap-root never thrive as well as trees grown by direct sowing.

The nuts are so attractive to rodents that direct sowing without special measures may be an expensive and total failure. In order to avoid losses to mice, sowing of the nuts stratified over winter is best done in spring. The suggested method of sowing with protection is as follows:

- Cultivate a metre-square planting site so that it is weed-free.
- Dig a hole 10cm deep and in the bottom insert a plastic 'quill' tube, pushing it in another 5cm or so (see page 186; use only quills with the weak seam for bio-degrading).
- Fill soil round the outside of the quill to

ground level and firm.
- Drop one nut down the tube, followed by soil to match ground level.

Alternatively, cell-grown plants 30cm high may be planted in treeshelters. Plant in mixture with ash.

Plant sites must be kept clean-weeded until the trees are growing vigorously. Walnuts are extremely susceptible to grass competition.

This is a specialised crop, like raising pedigree cattle, and the trees deserve careful tending. Their natural habit is to produce a wide-spreading crown. The aim in walnut culture should probably be to produce a fat butt log free of branches and defects about 3m long and, above that, a short log which then divides into two or three main branches. The butt log can be improved by pruning with secateurs. The second log, which carries the main break of the stem into branches, provides the interesting figure of 'flares' in the wood. The plantation, whether pure or with nurse trees, must be respaced and thinned from an early age to allow the selected walnuts to develop large crowns above the break of the bole, say at 4.5m, avoiding checks in development arising from a delay in opening. It would probably be a mistake for a grower to be greedy and try for extra length of clean bole by keeping the plantation 'tight' for additional years; the crown may never recover its full vigour.

Walnut timber is already valuable. Veneer-quality logs in Continental Europe fetch about £1,200 per cubic metre in the round (1997), three times best oak and 25 times construction-grade conifers; logs with flares and burrs fetch much more. A short log for gunstocks (the bottom one metre of the stem) may sell for up to £1,800 per cubic metre. After sawing, straight planking now sells in Britain retail at around £2,000 per cubic metre, either home-grown or French.

CHRISTMAS TREES

Farmers, if they think about a tree crop at all, probably regard Christmas-tree production as the obvious starting point. After all, suitable seedlings are available from nurserymen at around £100 per 1,000, ten pence each, and two-metre trees may retail in the city at £15; since one could plant 6,000 to 10,000 trees per hectare and a two-metre tree might grow in six years, it may appear to be a lazy way to make a fortune. If only it were as easy as that. Customers are choosy and demanding; lop-sided trees, yellowish foliage, lanky stems, bald patches in the crown and other irregularities are all magnified into fatal defects, particularly by the buyers on behalf of supermarkets. Nationally there is intense competition and considerable oversupply. Blindingly obvious is the fact that the demand is highly seasonal. In some places at least, theft is a serious problem.

This chapter will not guide the person who is contemplating becoming a commercial grower of Christmas trees; such a move would have to be prefaced by a thorough market survey and trials to establish the viability of investment in such an enterprise in the particular locality. Nevertheless, there are two possible levels of interest for the small grower: the modest production of specialist and high-value trees for a local market and the discriminating buyer; and the incidental production of Christmas trees planted as nurse trees or fillers in planting a new wood. The crop is quite compatible with game-bird shooting.

There have recently been some changes apparent in the Christmas-tree market which deserve consideration. For about a century, the traditional Christmas tree of Britain has been the Norway spruce. Freshly cut, taken into the house a few days before Christmas and removed on Twelfth Night, the tree in the first half of the century required a useful life of about 14 days in houses which were heated by open fires; a tree set in a window or an entrance hall had a cool environment (by today's standards, a cold one). In this regime Norway spruce was just satisfactory: by Twelfth Night it was beginning to shed needles.

Today it is usual for trees to be bought and decorated much earlier, ten days or more before Christmas, and the house environment, with central heating, is much warmer and markedly drier. In these circumstances (even discounting the fact that the commercial grower may have to cut trees in mid-November to meet the requirements of retailers so that they may 'catch the market'), Norway spruce may have shed its needles by Christmas Day, so that it both looks scruffy and makes itself severely unpopular with whoever vacuums the house!

There are chemicals which can retard the loss of needles somewhat and the buyer may also delay shedding by standing the cut stem in water. These make only marginal improvements, however, and the answer to the problems may lie in finding other kinds of trees. Whatever species is chosen it must retain its needles for more than three weeks, have non-prickly foliage and branches strong enough to hold ornaments; a pleasant resinous smell is an advantage.

Two genera provide good alternatives to spruce: silver firs and pines, both of which hold their needles even in a well-heated room and for the longer modern commercial Christmas-tide.

The European silver fir, the traditional tree in Germany, is unsuited to planting in Britain owing to insect attack, but the noble fir, *Abies procera*, is an attractive and hardy substitute, especially the form with the clearest blue-grey foliage (var. *glauca*). On the Continent the Caucasian silver fir, *Abies nordmanniana*, is also used, which has a colour and form close to that of the noble fir, although the seedlings cost double the amount. These are slower-growing trees, and the blue-grey or green foliage silver firs have come to fill the expensive top end of the market with a two-metre-tall cut tree in London recently retailing at up to £40 (£18 wholesale).

An obvious change in the Christmas-tree market in the last ten years has been the emerging acceptance of pines, both Scots and lodgepole. Scots pine is the most popular Christmas tree in the USA and Canada, in spite of the fact it is an exotic in a continent with an abundance of magnificent conifers. It has attained its premier position in sales as a result of growers providing a well-tended product which accurately meets the requirements of the customers. The tree holds its needles extremely well, is a good colour, has an attractive resinous smell and can be grown in a densely foliaged conical shape by pruning. The species is well accepted in the British market and now retails for more than Norway spruce in some parts of the country.

The second pine to appear on the market in recent years is lodgepole, *Pinus contorta*. The bright green colour is markedly different from that of Scots pine, but the tree holds its needles just as well. Some provenances of the tree tend to be tall and lanky with rather long distances between the branch whorls relative to branch length, so that the crown is, in the opinion of many people, rather too open. These long annual shoots are exaggerated when the tree is grown on fertile soils, and, while trees cut in respacing and early thinning may be marketed opportunistically, lodgepole appears to have little to justify using it in place of Scots pine for a commercial Christmas-tree enterprise. On the other hand, there has been almost no effort to cultivate and trim the tree as the Scots is in North American tree farms, and perhaps the system is waiting for develop-

ment, beginning by selecting suitable, less-lanky provenances.

Irrespective of the species being grown, the Christmas tree – if it is to be successfully marketed – must have a straight stem and a densely foliaged, compact and symmetrical crown. The main demand is for trees about 1.5–2m tall; they should have a basal 'handle' of trimmed stem about 15cm. Wide-spreading branches are inconvenient, so planting may be dense; Norway spruce may be planted as close as 90cm, pines at 1m, firs at 1.25m.

Commercial growers trim the crowns of spruce, pine and silver firs to achieve symmetry and dense foliage; hedge shears are the best tool. The objective is to trim the tips of the side shoots in late summer, inducing new bud formation and additional shoots. Shearing begins when the trees are about 90cm tall and continues annually to the year before harvest.

When the established tree's growth begins to accelerate, perhaps at about year four, so that it would produce a leading shoot too long for a high-quality tree, the removal of basal branches not only creates the clean 'handle' later needed at harvest but reduces the photosynthetic area and thus the length of the following year's leading shoot. Fertilisers should be used sparingly, if at all, and certainly not enough to cause fast, tall growth, but only to achieve good colour and abundant buds and foliage.

When consideration is being given to growing Christmas trees as a commercial venture, even on a modest scale, careful thought should be given to the site factors of the area, and it seems indisputable that the product should be aimed at the expensive and quality end of the market or for a strictly local one.

Soil: A sandy loam is probably best, but certainly a soil which has good drainage and moderate fertility. Too rich a soil will cause the trees to grow too quickly to height. Heavy clay and areas subject to flooding should be avoided.

Climate: Sites which are frost-hollows should be avoided. Places subject to severe winds are unsuitable, since the winds may cause the loss of leading shoots. Symmetry and undamaged leaders are essential.

Shelter: Although shelter from severe wind-storms is desirable, too close shelter with overhanging branches of surrounding trees will cause poor growth.

Access and Theft: It is helpful for marketing if the site has good car access in order that customers may come to choose their tree, as is now becoming popular in parts of Britain (it is almost *de rigueur* in the USA). Good access for customers, however, may also imply good access for thieves, so secure gating and fencing may have to be provided. Unprotected sites alongside public roads should probably be avoided, as they provide too strong a temptation and are too open to trouble.

HOLLY

The Romans gave twigs of holly for good fortune along with gifts to their best friends, and the Druids valued holly as the tree which, since it was evergreen, was never abandoned by the sun, which they worshipped. Now, like the Christmas tree, it is marketed for decorating houses at mid-winter; throughout the country there is a brisk demand and the retail price for branches is high, especially for those with berries.

This is essentially a crop for the small woodland grower, particularly for the woodland edge and perhaps in conjunction with a wildlife wood. The tree thrives best on a well-drained sandy loam soil, preferably slightly acid. Frost-hollows should be avoided, as should sites subject to severe winter exposure. For commercial cropping, holly should be spaced not less than 8m apart. There are many garden cultivars, some with variegated foliage, others with pale-coloured berries; the market requires moderately sized, dark green leaves and bright red berries. It is expensive and inconvenient to cut twigs from tall trees, and high branches carry leaves without prickles, so the plants should be topped and encouraged to grow to width. Harvesting should begin at about year seven.

Holly is a slow starter, but once established it grows vigorously. Plants may be raised from seed but it is best for the commercial grower to raise them from vigorous hardwood cuttings, which ensures the progeny are of the correct variety and the correct sex. Protect young plants from deer and sheep. In a plantation, one in ten or 12 trees should be a male, perhaps one in six if they are planted in a single line. The flowers in May and June are insect-pollinated and a commercial grower with a plantation should consider arranging for a bee-keeper to bring in a hive to ensure good fruiting. In order to maintain annual berry production and vigorous shoot growth, established bushes should be given a balanced N-P-K 10-10-10 fertiliser annually at the rate of about 0.5kg per tree. The shade of the tree itself should suppress all grass growth under the crown, but it will benefit from the application of an organic mulch in the root-feeding area just beyond the line of the lowest branches.

The produce may be sold to wholesalers, but, with a little care, it readily lends itself to retailing, by tying into attractive bunches and packaging. The life of the trees is heavily dependent on the way the produce is cut; ruthless cutting will quickly destroy the tree, which is why the job is best done by the grower, who has most interest in sustaining the crop, and not by a contractor or a casual worker.

GREENERY

In Britain there is a modest but steady market for greenery which florists use for wreaths and flower arranging, and there is now a brisk export to Continental Europe. Lawson cypress is the most common plant used. On the Continent, and particularly in Germany, the demand is very large and prices are high. It is regular practice to thatch grave plots in cemeteries, especially with silver fir foliage.

The long growing season for conifers in the maritime climate of the British Isles gives certain advantages for growing greenery, and small growers might readily develop this market, especially with the easier transport by rail to the Continent. Growers should concentrate particularly on high-value material such as *Abies procera var. glauca* and *Abies nordmanniana*, although cypresses and western red cedar are reliable. Trees should be planted at wide spacing, say 3m in the rows and 5m between rows, to allow a small tractor and trailer between rows. Grass immediately round the plants should be controlled with herbicides and the inter-row area should be mown at least once annually. After establishment the trees should be given nitrogen fertiliser regularly to encourage vegetative growth. Pruning of the foliage may begin at about year eight and continue for about nine years, when they should be pruned to death and replaced. Complete fence protection against deer and domestic animals is essential.

With this and all special products, the advice must be to make enquiries about markets before committing yourself as a grower. For some, the markets and practices are well developed and advice is easily obtained from the regular agriculture extension services and forestry consultants. For others – holly, perhaps – there is more scope for innovation and advice is scarcer. Remember, too, that environmental conditions vary widely; what is feasible in Sussex may be quite impossible in Sutherland.

FURTHER READING

Drake-Brockman, G.R. (1996), *Harvesting and Production of Woodfuel from Poplar Short-Rotation Forestry*, Forestry Commission Technical Note 18/96, Forestry Commission and ETSU

Jobling, J. (1990), *Poplars for Wood Production and Amenity*, Forestry Commission Bulletin 92, HMSO, London

Tabbush, P. and Parfitt, R. (1996), *Poplar and Willow Clones for Short-Rotation Coppice*, Research Information Note 278, Forestry Commission, Edinburgh

Taking Stock

15

An incident during a forestry society meeting in Scotland remains vividly in my mind, and probably in the minds of others who were present. The owner of the famous host estate showed us Scots pine cut in thinning an excellent plantation; proudly he told us that the timber merchant, who was present in the party, always paid a premium price for trees from this wood and he mentioned the price paid, which was indeed slightly above average. Quite innocently one of the visitors then asked the average volume of the trees cut, which lay in front of us. Promptly the owner gave the answer, amid gasps of incredulity from the professional foresters present, for the merchant (who was gazing at his boots) had persuaded the owner the volume was about half the true figure. The memory of the incident explains why this chapter has been written.

What farmer would boast of getting the top price at market for his lambs, when in reality 100 lambs had been taken from the farm but only 55 paid for?

Professional foresters devote great care to the measurement of timber volumes and growth rates, the complexities of which would be out of place here. What is appropriate instead is to give the owner of small woods or single trees confidence that he knows what he is selling and is not being deceived outrageously. The methods described are not accurate but can give good guidance.

To make measurements of trees and timber, two items might be bought and some simple ones could be home-made. From a forestry equipment supplier buy a metric tape, 25m long, marked in 10cm intervals, and a metric 'diameter' tape which is marked in centimetres. (Lest there is confusion, this tape, when put round the girth of a tree, allows the user to read off the equivalent *diameter* directly.) Also needed is a pocket rule measuring in centimetres and millimetres; most households or home workshops have one. From the nature of the instruments it will be obvious that the forest industry operates in the metric system (it has done so since 1971).

For measuring tree heights, the first item required is a straight lath of wood about 0.5m

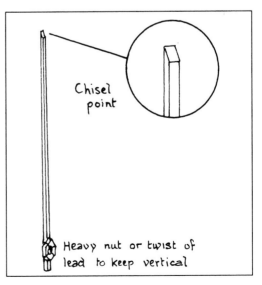

Chisel point

Heavy nut or twist of lead to keep vertical

MEASURING LATH

long x 1.5cm x 1cm, bluntly chisel-pointed at one end and weighted at the other, perhaps with a twist of lead or a heavy nut. To work with it is a staff conveniently 2m long, treated so that the few centimetres at both the bottom and the top can be clearly identified from a distance, generally by painting them a brilliant colour. The staff should be robust enough for woodland work but not so heavy as to be a burden. For measuring very tall trees, say over 25m, a 3m or 4m staff is preferable, but it then becomes cumbersome to carry.

An optical hypsometer could be purchased for really accurate measurement, but it is an expensive instrument which could be justified only by a heavy work programme, unlikely in small woodlands.

MEASURING LENGTH

With the normal tape, straight tree lengths and log lengths are measured in metres, rounding down to the nearest tenth of a metre up to 10m and to the nearest whole metre above 10m.

MEASURING TREE HEIGHTS

The height of a standing tree can be measured using the wooden lath with the weighted end and the long staff.

1. Place the 2m staff against the tree to be measured and find a position, about a tree height distant, where you can see the top and bottom of both tree and staff.
2. Hold the plumb-bob lath between finger and thumb near the top so that it hangs vertically at full arm's length, and aim the tip of the lath at the top of the staff. Then work finger and thumb gradually up or down the lath, still at arm's length, until the top of the thumbnail coincides exactly with the bottom of the staff (the tip of the lath still coinciding with the top).
3. Keeping finger and thumb steady on the

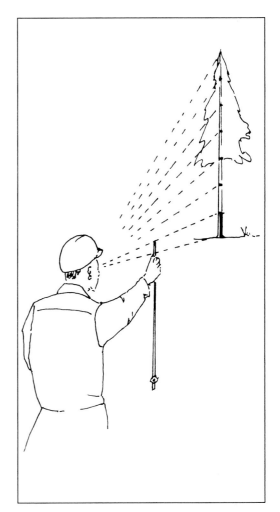

HEIGHT MEASURING
With the arm outstretched and the lath hanging vertical, measure the length of the known staff against the tree bole and then 'step' up the tree.

lath and letting the plumb-bob hang vertically, measure up the tree in 'steps', carefully noting where the tip of the lath comes each time on the tree stem. The number of steps to the top times the height of the staff is the height of the tree.

The method is useful in that the distance to the tree need not be measured (no tape, so you can work single-handedly) and it works

Make sure you are looking at the true top of the tree — otherwise the height will be wrong.

MEASURING HEIGHT OF CONIFERS AND BROADLEAVES

the number of steps. The height to be measured may be to the top of the tree or to a major fork or break in the stem which will be the limit of sawlogs. The method does not measure height accurately (for that you need a hypsometer) but it gives a fair approximation, sufficient for interest and to avoid being taken advantage of.

In conifers the top of the tree is usually easy to identify, since most species have a spire-top. In mature broadleaves, however, the top of the tree is usually rounded and a side branch may easily be confused with the true top, especially from relatively close; the result is an overestimation of height.

Measuring the top height of young trees of known age is used in calculating the potential productivity of stands. The height growth of conifers is virtually independent of their stocking; isolated or crowded (within reason), their height-for-age is the same, and over many years this relationship has been correlated with the volume production achieved by well-managed stands. The results, for all the main species, are yield tables published by the Forestry Commission, which can be used to predict volume growth and to guide management.

MEASURING DIAMETERS

The diameters on the ends of cut logs may be measured directly with a straight ruler in centimetres, rounded down to the nearest centimetre. The cross-cut stems show that, although nearly circular, tree boles are usually somewhat elliptical, since wind pressure causes one side of the bole to grow more than others (as indeed the weight of a branch affects its cross-section, the elliptical shape forming a more efficient structure). An average of two diameters may be taken.

The diameters at the middle of logs and the boles of standing trees cannot be measured with a straight ruler; they may be measured

well on sloping ground, when you should work from the uphill side of the tree. As the angle of sight towards the top of the tree becomes acute, accuracy is lost, so get well back from tall trees. The more 'steps' that are involved, the greater the inaccuracy, because successive steps may not begin exactly where the previous one ended. A staff longer than 2m may be worth using, in spite of its greater inconvenience in handling, in order to reduce

with callipers but the usual method is to measure round the girth with a tape calibrated in centimetres diameter. The relationship between girth and diameter is:

Circumference of a circle (i.e. girth) = π x diameter

so $\qquad$ Diameter = $\dfrac{\text{girth}}{\pi}$

(π being the constant 3.142)

Using a tape round the girth automatically averages out the differences of diameter due to elliptical growth and the diameter tape makes the above calculation automatically. When using a tape, make certain it is not twisted and that it goes round the stem at right angles to the axis, avoiding bumps and branches.

When measuring the basal diameter of standing trees the convention is to do so at 1.3m above ground level, so called 'breast height'; this is taken as the basal size of the tree, to avoid the root swellings and buttresses at the base itself. The diameter at breast height is abbreviated as 'dbh'. On sloping ground the measurement to 1.3m is made on the upper side of the tree. Conventionally trees less than 7cm dbh are classed as unmeasurable. If the stem is swollen at 1.3m, measure above the swelling. If a tree bole divides into two at or below 1.3m, count it as two trees and measure each above the fork.

The measurements are made over the bark; if there is any possibility of confusion or doubt, the figure is given as so many centimetres over-bark (cm ob).

You may require to know the average dbh of a substantial number of trees (for instance the trees marked for thinning); whether you measure them all or a sample, remember that the average tree can be found only be adding together the *cross-sectional areas* – not the simple diameters – and dividing by the number of trees, which is considerable trouble. A working approximation which

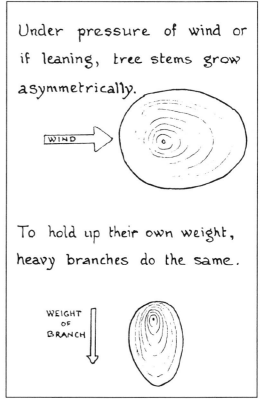

ELLIPTICAL TREES

avoids that complication can be obtained by using the *40 per cent rule*.

Using a diameter tape, measure the dbh of all the stems or those of a large sample, say 80 trees, and record the numbers as shown in the enumeration diagram. The sample must not be selected; you should walk through the wood on a set line across any obvious gradient or variation in the stand, taking every tree you come to without selection or rejection. Whatever is the total number recorded (80 stems in the example), calculate 40 per cent of it (32), and count back on the form from the largest recorded tree. The class of the 32nd tree (in this example), counted back from the largest, is the size of the average tree (11cm in the diagram). The volume of the average tree may then be established by cutting and accurately measuring two or three of that size, and multiplying the average volume by the total number of trees in the wood or in the thinning. You will then know how many cubic metres you are sending to market.

Whether the average tree is calculated by the 40 per cent rule or by a more exact method, you should beware of lumping together two (or more) groups of trees which are really separate. These might be two species, say thinning a mixed plantation of pine and larch, or they might be two areas of the wood which, although the same species, have grown at very different rates because of soil difference, a situation easily recognised on the record sheet by the existence of two peaks of frequency. In all such cases, keep the records of the groups separate and calculate separate average trees.

In small woodlands it is often more convenient and appropriate to count and measure the dbh of all the trees for thinning or felling rather than take a sample. If you do decide to measure by sampling, a square plot 10m x 10m or a circular one of radius 5.64m contains one-hundredth of a hectare, which is easy to relate to a known total area. But beware! Any bias or inaccuracy in the plot will be multiplied many times when the total is calculated, so if you use sample plots, measure several in order to reduce the errors of bias, and be aware of obvious anomalies such as edge trees.

MEASURING VOLUME

The unit of volume is the cubic metre (m³) taken to two decimal places if appropriate. The minimum top diameter for volume measurement is generally 7cm ob, or the point at which the main stem is no longer distinguishable because of division into branches.

The volume of a tree bole after felling can be measured by marking the stem off in

ENUMERATION		
COMPARTMENT..4...		DATE 21/11/95.
DIAM. CLASS CM. ob		TOTALS
8	𝍸𝍸 𝍸𝍸	10
9	𝍸𝍸 𝍸𝍸 𝍸𝍸 I	16
10	𝍸𝍸 𝍸𝍸 IIII	14
11	𝍸𝍸 𝍸𝍸 I	11
12	𝍸𝍸 𝍸𝍸 I	11
13	𝍸𝍸 IIII	9
14	𝍸𝍸 I	6
15	II	2
16	I	1
	TOTAL	80

ENUMERATION AND CALCULATING THE AVERAGE TREE

Total number of trees = 80
40% of 80 = 32
Count back from the largest:
1 + 2 + 6 + 9 + 11 = 29 (Classes 16 to 12 incl.)
+ 3 of Class 11 = 32
Therefore the average tree of this enumeration is in Class 11.

sections and, with a diameter tape, measuring the diameter at the midpoint of each section. Then calculate the area of the cross-section in square metres (area = πr^2, where r is the radius, i.e. half the diameter) and multiply by the length of the section in metres. Addition of the sections gives the volume of the tree. If the sections are longer than 6m, the result is apt to be inaccurate, because the half-length diameter is liable to be not truly representative of a long length of bole. A good practice is to measure the bole in 5m sections. In conifers only the volume of the main stem to 7cm is calculated and branches are ignored.

To calculate the area of the cross-section, measure the diameter in centimetres; divide by 2 to get the radius and square it; multiply by 3.142 and divide by 10,000 to convert square centimetres to square metres. For example, if the diameter is 18cm:

radius	= 9cm
radius squared	= 81cm²
multiply by π	= 81 x 3.142
	= 254.5cm²
divide by 10,000	= 0.025m²

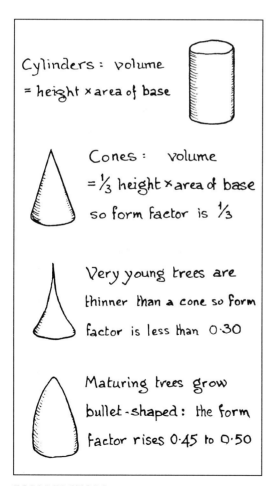

FORM FACTORS

Coniferous trees, which are simpler to measure than broadleaves, have a shape which approximates to a very tall, thin cone. The shape is not constant through the life of the tree; when young it is more drawn up than a true cone, whereas in late life it becomes bulkier in the middle, somewhat bullet-shaped. This offers another method of estimating the volume of standing trees: to apply a 'form factor', as follows:

1. Measure the diameter at breast height in centimetres and calculate the stem's basal area in square metres with the formula area = πr^2.
2. Measure the total height of the tree in metres and multiply by the basal area; in effect this is the volume of a cylinder, the tree's height on its basal area.
3. Now multiply by the estimated form factor to change from a cylinder to a near-cone shape and give the estimated volume of the tree in m³; use a factor of 0.5 for mature conifers in a plantation, 0.4 for open-grown conifers and 0.35 for younger trees.

As a rough guide against your own calculations, a conifer of 10cm dbh may have a volume over-bark of about 0.03m³; 15cm dbh 0.08m³; 20cm dbh 0.23m³; 25cm dbh 0.45m³; 30cm dbh 0.7m³.

Broadleaves are more difficult to measure standing than conifers; height is usually taken to the break of the crown (where the main stem is no longer distinguishable and it divides into the main branches) and can be multiplied by the estimated cross-sectional area in m² at mid-length. After felling, the volumes of broadleaves are best treated by dividing the stem into straight sections (or actual logs) and calculating the volume of each by the mid-section area times the length. In a mature broadleaved tree the volume of the branchwood is often about the same as the volume of the sawlogs, less in younger trees.

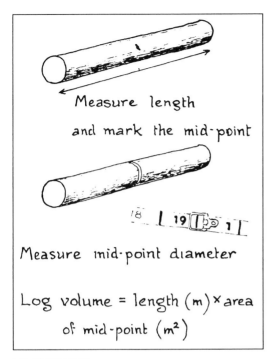

Measure length and mark the mid-point

Measure mid-point diameter

Log volume = length (m) × area of mid-point (m²)

LOG VOLUME

Logs for sawmilling and veneering may be measured by the method of mid-section area times the length, as already described. Some forest growers and the sawmillers dealing regularly with them now measure sawlogs by the top (small end) diameter rounded down in 2cm classes; the volume is the small-end area multiplied by the length, the argument being that the size at the small end defines what a log can be sawn into. In the case of sharply tapered trees, this method, although realistic in respect of sawnwood content, gives a volume considerably smaller than the mid-section measurement method and smaller than actuality. The unmeasured wood will be sawn away as slabwood and sold by the merchant as residue for pulping or particle-board.

Conifers grown in exposed conditions tend to be shorter in height for any given basal diameter than those grown in shelter and consequently the stems are more sharply tapered. As sawlogs or as transmission poles this is an unhelpful character but it is one which must be accepted.

BARK ALLOWANCES

In Britain all measurements of trees and logs are taken over-bark and this should be stated in any sale agreement. Bark thickness varies between species and with age, so no allowance can be generalised. If you are undertaking conversion of wood and manufacture of items for market, you should expect to lose about 10–15 per cent of the log volume as bark.

CORDWOOD

Coppice wood and the branches of broadleaves are often cut into lengths for firewood and piled for sale. Traditionally the lengths cut were four feet and they were stacked in a pile eight feet long and four feet high, known as a cord; this contained 128 cubic feet, wood and air. With metrication it is sensible to adopt the Continental European measure, cutting lengths of one metre and stacking one metre high and one metre wide. A cubic-metre-stacked measure contains approximately 0.6m³ solid wood.

SHORTWOOD

Coniferous wood intended for pulping and particle-board manufacture is usually cut in the forest to lengths of two or three metres and is sold by weight which is calculated by passing the transporting truck over a weighbridge at the gate of the manufacturing mill. The diameters will be random within tight specifications set by the purchaser, dictated by the capacity of the processing machinery.

The weight of wood varies according to its moisture content and with species, from wet oak, beech or hornbeam to dry spruce or pine, although the material of the cell walls of all species is approximately the same, about

1,560kg per m³ (a specific gravity of 1.56); it is the air spaces which account for the differences. In standing trees the sapwood has a higher moisture content than the heartwood and consequently is heavier. As a result, the weight per unit volume of poles (in which the percentage of sapwood is higher) is greater than that of large trees. For instance, freshly cut poles of Sitka spruce sink in water (i.e. weigh more than 1 tonne per m³), whereas large logs float. For working purposes, a solid cubic metre of hardwood soon after cutting may be taken to weigh 1 tonne (1,000kg) and one of softwood 750kg, but as the wood dries both these figures reduce.

POLES AND STAKES

Poles, stakes, round stobs, fence straining posts, bean-poles and similar products are normally sold by number of specified sizes. The same is true of transmission poles for electricity and telephones, sizes being closely defined by the purchaser.

YIELD CLASSES

In a normally stocked, even-aged plantation, the aggregate volume of stemwood produced up to date per hectare (standing volume plus thinnings), divided by the current age of the stand, gives a figure known as the mean annual increment (MAI) – the average volume growth per year throughout that stand's life to date. The unit is m³ per ha per annum. In the early life of a stand the MAI gradually increases, then it culminates at between 45 and 70 years of age in many conifers, and then it declines with age. The volume figure per hectare at which MAI culminates is referred to as the Yield Class (YC) of that particular stand; it has been shown that this figure can be predicted accurately from the height-age growth of young stands, which is useful for planning the work schedules and predicting future sales from large forest stands. Yield-class calculations require the forester to measure the top height of five or ten of the stoutest trees of the species in the stand; the average of these identifies the yield class from a published set of height/age graphs. For instance, in a Scots pine stand, if the top heights of the largest trees average 15m at age 30 years, the stand is probably YC14; if the average is 10m, the stand would be YC8. Yield classes may range from as low as 4 for many broadleaves to 24 or even more for some conifers on fertile sheltered sites. YC14 to YC18 is very satisfactory for Sitka spruce, Douglas fir, etc. YC8 to YC12 is a common range for Scots pine. YC8 or YC10 is normal for larches (range 4-14). YC4 to YC6 is normal for oak and beech (range 4-10).

For instance, a YC14 stand of a pine, cut at age 70 when its volume yield is at maximum, would produce 980m³ in all over the period (i.e. its maximum MAI of 14m³ times its age, 70), cut approximately half as thinnings and half as the final stand. If it is left to grow on beyond 70 years, it will, of course, produce a greater volume, but its annual growth will gradually slow down, as will its MAI. All this refers to volume, not value. Weighing the benefits of growing bigger trees against the costs of waiting extra time for the main harvest is part of the task of management, which is beyond simple arithmetic.

The yield-class measure is a useful guide to the long-term productivity of the woodland. If it is YC18, it means the woodland can produce a maximum of 18m³ per ha each year. It will average less than that if you leave it to grow beyond the rotation of maximum mean annual increment – say, to get the advantage of higher prices for extra large logs – or if you have to cut early because of windblow, but you cannot count on more than 18m³ per year in the long term. Knowing the yield class of the woodland is a useful guide.

Measuring the volume of trees can be interesting simply to discover how well they are growing and how big they are. Taking stock, however, has real significance in two respects. It is clearly important to know what amount of wood is being sold, the equivalent of the farmer knowing how many lambs or tonnes of grain are being sent to market. And it is also essential in a sustention-managed woodland to ensure that the harvest is continually kept within the yield – i.e. that the capital is not being cut into. That fault is the equivalent of a sheep farmer sending breeding ewes to market, a ploy which could provide a quick income but spells disaster in unsustainability.

TAKING STOCK OF AGE

An obvious and vital aspect of taking stock is the critical appraisal of woodlands which have been untended for some time, so that you may decide what needs to be done to put them in good order. You need to ask:

- What trees are here – species, size, age, condition?
- What is the future of this stand?

The answers provide the information for a plan of operations (chapter nine).

Taking time, and with eyes open, walk slowly through the wood and make careful written notes. It may be appropriate to enumerate the trees as in the diagram on page 255; it is useful to annotate a large-scale sketch map to show the main features (distinct stands, different species, perhaps individual trees if the woodland is very small).

The answers to the first question are factual; those to the second require judgement in the light of the long-term objective for the woodland. It is important that, in addition to describing the numbers and sizes of the trees, you should estimate the distribution of their approximate ages. Are all the trees very large? Apparently the same age? Are there no middle-sized, middle-aged of the main species? No poles, saplings, seedlings? Although their time scale is longer, woodlands are no different from a population of animals or human beings in that a community composed entirely of geriatrics has a strictly limited future, in contrast to one which has middle-aged, teenagers and children. Having taken stock of the age distribution, the next step is to consider what may be done to allow the woodland to fulfil its management and silvicultural objectives in a sustained way.

FURTHER READING

Edwards, P.N. (1983), *Timber Measurement – A Field Guide*, Forestry Commission Booklet 49, HMSO, London

Hamilton, G.J. (1985), *Forest Mensuration*, Forestry Commission Booklet 39, HMSO, London

Wood from the Woodlands

16

It would be out of place in this book to discuss the design of sawmills or the chemistry of paper-making. Nevertheless it is such end-uses of wood which fix its value. The seller of timber should understand what buyers are looking for and why certain specifications are set; knowledge of the important qualities and defects should help the grower to make his woodland financially productive by meeting the customer's requirements.

Wood is superbly versatile material and the demand for it keeps on increasing for an immense number of uses. We use it in the round, split, sawn, turned, chipped, sliced and pulped, as fence posts, flooring, furniture, sculpture, pea-sticks, for building construction, paper-making, fuel . . . the list is endless. Different tree species have different attributes, fitting them more for one use than another, and the way the individual tree has grown – its straightness, the size of the stem, the presence or absence of branches – affects its usefulness for particular purposes. All these affect the price wood fetches in the market. The pages of chapter three list some specific uses.

The trade and consumption of timber and wood products in the UK in 1994 is shown in Table 16.1, together with home-grown supply; this shows the low rate of self-sufficiency and the great dependence of the British Isles on imports. The figures also show the heavy demand for softwoods, i.e. conifers.

Table 16.1

WOOD SUPPLY AND CONSUMPTION, UK 1994

	Imports m^3	Exports m^3
Wood, sawn & in logs		
Softwoods (conifers)	16,800,000	300,000
Hardwoods	1,500,000	200,000
Wood-based panels	4,700,000	800,000
Paper & Board		
Paper	16,500,000	3,700,000
Pulp	8,200,000	100,000
Waste Paper	500,000	1,300,000
TOTAL	48,200,000	6,100,000
VALUE	£6,441 million	£2,287 million
NET TRADE DEFICIT	£4,154 MILLION	

UK HOME-GROWN WOOD PRODUCTION

TOTAL 8,050,000 m³ of which:

Softwood sawlogs	50%
Hardwood sawlogs	4%
For wood-based panels	20%
For paper & paperboard	18%
Other industrial wood	5%
Fuelwood	3%

Note: The volumes of panels, paper and board are the wood raw-material equivalent – the volume of round-wood required to make the manufactured goods.

Figures from Forestry Industry Yearbook 1995, published by Forestry Council of Great Britain

In order to understand the values and defects of different trees and why wood is processed in particular ways, it is useful to recall the basic structure of the typical tree stem, described in chapter two.

In the living tree the wood performs three functions: holds the tree up, carries sap and stores products. The sapwood carries sap upward from the roots to the crown and serves in food storage, which makes it a target for insects and fungi, if they get the chance to attack. Sapwood is light-coloured and non-durable. Heartwood is often darker-coloured, from the tree's deposition of by-products such as tannins and oils; in some species it is naturally resistant to decay.

Chemically wood is made up of cellulose, hemi-cellulose, lignin, extractive chemicals and ash. Cellulose and hemi-cellulose make up over 60 per cent, being the main constituents of the fibres and cell-thickening; they are a form (a *polymer*) of glucose and other sugars, chains of sugar units.

Lignin comprises about 30 per cent of softwoods and rather less of hardwoods. It is a complex substance, like a natural plastic, which serves to glue the cells together and give them rigidity. Chemically it is relatively insoluble, and in many processes using the cellulose of the wood (as in some kinds of paper-making) it is necessary to remove it, although difficult to do so.

The extractive chemicals in wood include tannins, essential oils, fats, resins, gums, acids and starch. They may make up to 10 per cent, depending on the species and the time of year, not forming part of the structure of wood but contributing importantly to its colour, resistance to decay, smell and how it behaves in use. The chemicals in some woods are irritants; among British timbers box and yew may cause skin, eye and respiratory inflammation, especially as dust from sandpapering and machining. Some tropical timbers are much worse.

Ash-forming minerals generally comprise no more than 2 per cent of wood (and sometimes much less). Calcium, potassium, phosphates and silica are the most common.

SAWLOGS

In the common conifers, the softwoods, logs with a top-end diameter of 14cm are potentially sawmill material but there are important conditions. Logs about 5m in length and 20cm in top diameter may be cut for building construction; small sawlogs of 14cm and up may make pallet wood, etc. Obviously, the log must be at least as long as the end product, so logs to be cut into house joists and rafters, for instance, must be at least 4.5m long, and similarly for other markets. Above all, logs for sawing must be straight;

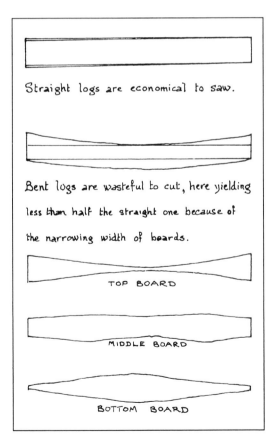

Straight logs are economical to saw.

Bent logs are wasteful to cut, here yielding less than half the straight one because of the narrowing width of boards.

TOP BOARD

MIDDLE BOARD

BOTTOM BOARD

A BENT LOG
Straightness is the prime requirement for sawlogs.

not only are crooked logs wasteful to saw into straight boards, but the sawn battens are of doubtful quality, because of the different directions the grain must run in them.

Aim to market sawlogs; the price differential over pulpwood is large. Apart from stems of exceptional quality, timber merchants want to be assured of a truckload of saw timber as a minimum purchase, say 30m³.

Given logs are sufficiently long and straight, larger-diameter ones are more useful than smaller, up to a point. There is less waste in cutting large logs, there is proportionately less weak juvenile wood and there is generally a size premium for large-dimension timbers. Modern sawmills, however, are normally designed to handle logs of a limited bracket of sizes and a manager would not wish to buy any trees larger than those for which his sawmill has been designed. This fact distorts the timber market locally, and probably does so nationally as well.

The price-size diagram, which is based on large numbers of measured sales, shows, as a national average, the relationship between the volume of a tree and its value per m³ when sold standing in the forest. The upper curve implies that conifers less than about 13cm in diameter at breast height (dbh) have no market value in a standing sale (1993

prices) but above that, up to, say, 25cm dbh, the value rises steeply to about £20 per m³; thereafter the value rises rather more gently to about £36 per m³ at an average of 35cm dbh; but above that size the value rises little, to less than £40 per m³ at a tree size of about 55cm dbh.

These are average figures, useful as a general guide in economic planning but not for calculating the outcome of a particular sale. Superficially they appear to deny that

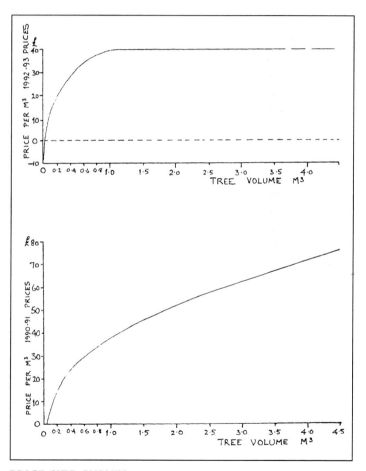

PRICE-SIZE CURVES

Top: Price-size curve for conifers, UK.

Below: Price-size curve for oak and ash, UK.

(After Forestry Commission Bulletin 68, Occasional Paper 32 and Research Information Note 226)

large logs are more useful than small ones; the explanation is that sawmillers in Britain have not found sufficient large logs on the market to warrant investing in large saws, and to that extent any price-size curve is a self-fulfilling prophecy. The curves are considerably affected by the haulage distance from the woodland to the mills. For instance, the price-size curve for conifers in Scotland shows that average prices for standing timber are generally 20 per cent lower than in England and Wales, largely a result of longer road-haulage distance to major markets.

The fact that these price-size curves are a product of the design of local mills (and hence their ability to use logs only up to the capacity of the machinery installed) is confirmed by the equivalent curves for other countries. In western Canada and the USA, for instance, prices flatten out at a tree volume around 20m³ (say 1.0m dbh). Where there are sufficient large logs to make it worth investing in a mill to process large logs, merchants want to take advantage of their higher efficiency and are prepared to pay extra for them.

These points are also confirmed by the shape of the price-size curve for broadleaves, of which one is shown here for oak and ash, based on 1990–91 prices. All the curves for the major broadleaves show prices still steadily increasing at 4.5m³ per tree (the top limit at which data were available from the survey). Sawmills and veneer mills for broadleaves have not been designed down to a particular size of log and therefore the extra usefulness of the large-size logs is reflected in their prices; there is a market incentive to grow large broadleaves.

The saws used for converting logs may be bandsaws, circular saws or reciprocating frame saws. Whatever the design of the machine, the maximum amount of sawn full-length boards and battens that can be cut from a log is set by the size at the small end, the top of the log. A

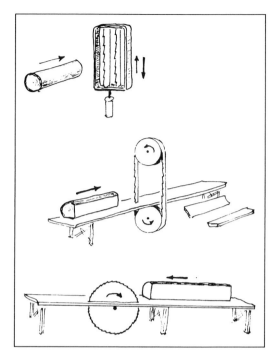

TYPES OF SAW

Top: the Frame saw reciprocates with (usually) several thin saws held in tension in the heavy frame, preset to cut a given size of log to customers' orders. It thrives on long runs, not versatility.

Middle: Bandsaw; a motor drives the lower wheel while the top one maintains tension on the saw. Very large logs can be cut: very versatile, so mixed log sizes can be cut.

Bottom: Circular saw; size of cut is limited to less than half the saw diameter.

sharply tapered log contains much more wood than a truly cylindrical one of the same small-end size, but the extra can be used only for residue markets (for paper-pulp, particle-board, etc.) or by costly rehandling and resawing of the slabs, the curved pieces sawn from the outside of the log.

Some modern sawmills avoid sawing the slabs from the round logs by using a 'chipper headrig', which, at incredible speed, chips away the outside of each log to produce a square section balk which is then sent to the saws for cutting into battens. Whether the conversion produces slabwood or chips from a chipper headrig, everything which is not to be sold as sawn lumber is nowadays sold as 'residue', mainly for manufacture into particle-board or paper-pulp. That includes sawdust and off-cuts and it may include bark for some processing, although the trend is to remove bark from the logs as they arrive at the mill and use it for burning or for sale as a soil mulch.

Saws and chipper headrigs are machines run at high speed, a bandsaw at about 40m per second, and they are designed to cut wood. Consequently, if the log contains foreign bodies such as nails or pieces of fencing wire, there are devastating results when the saw encounters them. The threat of the ensuing costs of repairs and lost production is a major reason for timber merchants regarding farm-grown timber with suspicion and offering low prices. It is of the greatest importance that proprietors of small woodlands should exercise care to ensure that the logs they sell are not downgraded by the presence of wire, nails or other material buried in them. Never fix notices or fence-wires to trees.

Tree species best serve different end-uses. Spruces, pines and Douglas fir go to construction timbers (joists, rafters, studs), joinery timbers (flooring, sarking, carpentry), manufacture of pallets, etc. Larches are used

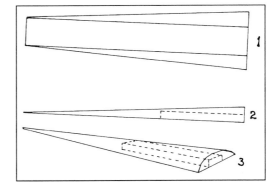

LOG TAPER
1. Top: For sawmilling, the usefulness of a log is set by its top diameter.
2. The slabs cut from a sharply tapered log may appear useful but, for what may be the product, there is a lot of extra work.
3. The curvature of the slab severely limits what can be cut. Nowadays most sawmill slabs go as mill residue for particle-board or paper.

for outdoor woodwork and specialist purposes such as boat-building and whisky vats, where the wood is kept permanently moist; when dried, larch timber has a tendency to split and twist. Beech is best used for furniture, especially chairs; sycamore makes excellent flooring; oak is best used for heavy construction and furniture; and so on.

Conifer logs over 14cm in top diameter are potentially sawlogs, if they are straight. In order to yield timbers for building construction, they must be around 4.5–6m long and, it is worth repeating, straightness is paramount. Rotten hearts, shakes and rotten or 'powder' knots are serious defects, because they affect the strength of the sawn battens and also the amount of saleable sawn wood which can be cut. The sawlog specifications in general use in Britain are given in the Forestry Commission's Field Book 9.

Pine logs deteriorate quickly, especially in summer, owing to a fungus which stains the wood pale blue. Sell pine quickly after felling

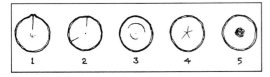

LOG DEFECTS
1. Frost crack
2. Drought cracks or drying checks
3. Cup shakes, following annual rings
4. Star shake
5. Heart rot

or windblow if this degrade is not to reduce the price. Storing logs in water guards against blue-stain fungus.

Sawn material may be visually graded; a grader will downgrade battens with large knots, those undersized or those with very wide growth rings, on the score that they would prove to be too weak for their purpose in construction. Visual grading is still employed but in modern sawmills it is now usual for battens which are to be sold to the building industry to be passed through a stress-grading machine. As the batten moves over its rollers, the machine subjects the wood to a bending force and measures its actual deflection. This proves whether materials meet or fail the specifications of architects and engineers for a particular job.

The common causes of failure of conifer timbers in machine stress-grading (and indeed in visual grading) are juvenile wood and bad knots. The wood produced in that part of the stem which carries green branches, the inner annual rings of any log, tend to be wide and composed of thin-walled cells. Such wood is markedly weaker than that formed later and it causes the sawn battens to be weak. Since the tendency is now for conifers to be grown on shorter rotations than formerly, the years when there are green branches on the lower bole and when juvenile wood is formed make up a substantial part of the production period, and consequently juvenile wood forms a significant part of the total log. Juvenile wood in conifers has become more important than previously. In the past, the known weakness of juvenile wood was the reason for sawyers, when converting large logs, to 'box the heart', i.e. to cut out the central part of the log for a relatively undemanding use. With small-diameter sawlogs that is not possible, and the weakness must be accepted. The minimisation of the juvenile wood in the valuable butt log is the principal reason for

the recent revision of forestry practice to plant Sitka spruce at not more than two-metre spacing.

The width of growth rings outside the zone of juvenile wood is not critical until the rings become exceptionally wide. A good general rule is that conifer growth up to about Yield Class 18 is acceptable in that this itself is unlikely to cause downgrading in stress-grading machines.

Ideally hardwood logs are over 3m long, straight, over 45cm in diameter, cylindrical with the heart in the centre, clear of knots, epicormic shoots or spiral growth and are winter-felled; few attain this standard.

SEASONING

Green timber contains a large amount of free water in the cell cavities (the vessels etc.) and the cell walls also are saturated. Drying out most of this water constitutes *seasoning*. When all the free water has gone (*fibre saturation point*), the cell walls begin to dry and shrinkage begins; if the bound water is removed too quickly, the cell walls may break and the wood will distort.

Shrinkage on drying is not uniform; longitudinal shrinkage is only about 0.1 per cent to 0.3 per cent, so 5m battens need be cut only about 10mm longer when green. Shrinkage is greatest tangentially to the surface of the bole, and in a radial direction about half that amount. Tangential shrinkage of wood dried to 20 per cent moisture content is generally 2.5 per cent (it varies with species) while radial shrinkage is only 1.5 per cent (5.5 per cent and 3.5 per cent when dried to 6 per cent moisture content). The effect of the girth of the log shrinking more than the radius explains why cracks develop when a disc is cut from the end of a log. ('Moisture content' means the weight of moisture in a piece of timber expressed as a percentage of the dry weight of that piece.)

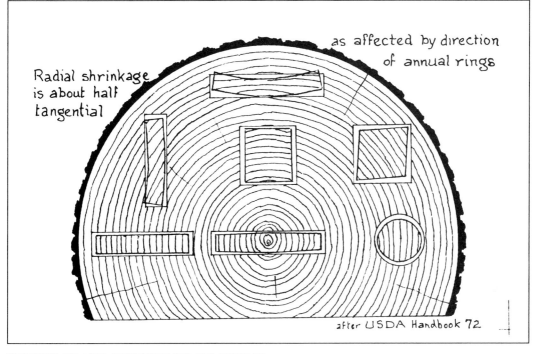

Radial shrinkage is about half tangential

as affected by direction of annual rings

after USDA Handbook 72

SHRINKAGE AND DISTORTION ON DRYING

The differential shrinkage is important for the sawmiller. Not only must he cut 'green' planks some 5 per cent larger than the required dry size, but the planks tend to distort during seasoning. In a plank the tangential shrinkage is greatest along the longest complete annual ring, so that it distorts into a shallow cup shape. Planks which are radially cut ('quarter-sawn') shrink but seldom distort. Distortion can be minimised by careful stacking, so that air circulates around the planks but there is pressure to keep them flat; stack on a firm, flat base and separate each layer of boards from those above and below by inserting wooden 'stickers' (seasoned softwood 25mm x 25mm) so placed that one is directly above another, supporting each board without any sagging.

By air drying under cover the moisture content of wood can be reduced to around 14 per cent, sufficient for outdoor work and general carpentry but too high for use in a centrally heated house, for which 10 per cent is a maximum and 8 per cent or less is better. In modern practice this is achieved by kiln-seasoning. The wood, either green or air-dry, is stacked with stickers in a closed kiln in which the air is circulated by fan, with the temperature and moisture content fully controlled; seasoning can be complete in a few days or even hours.

Fully seasoned wood is not only free from further shrinkage and cracking in its final use as furniture or construction timbers but is much less liable to attack by fungi and many wood-boring insects.

KNOTS

In spruce and Douglas fir, the knots of side branches tend to be well grown into the wood, and downgrading due to the presence of knots of normal size is rare. In pines,

Through·and·through
—quick and easy but most
planks are tangentially cut

Quarter·sawn
—more work but more planks
are radially cut & more valuable

SAWMILLING

however, two factors are important: branches are in tight whorls and hence knots are clustered, and there is a tendency for the dead branches to produce encased knots or powder knots which are not 'grown-in' to the wood. As a result, downgrading of pine timber on account of knots is more common than in spruce and Douglas fir. The owner of small woods should seriously consider pruning pines as a means of producing high-value timber, which the big-time forestry grower cannot even attempt. Begin pruning early; the diagram of knots in chapter six (page 145) clearly shows why.

For the more exacting jobs of joinery finishing, the carpenter and cabinet-maker want clear softwood timber (knot-free), but very little such wood is available. Since most conifers in Britain have been grown at rather wide spacing and have not been pruned, they are rather heavily branched and the wood is knotty. The price differential is considerable: sawn Scots pine with knots costs about £280 per m^3, whereas similar wood in 'clears' costs around £1,100 per m^3 (1996 wholesale prices).

CLASSIFICATION AND PRESENTATION OF HARDWOOD SAWLOGS

The specifications and uses of hardwood sawlogs are much more varied than for softwoods, and prices also vary more widely. Since few broadleaves are now used for construction timbers, long lengths are not so important as in conifers, but quality has a marked effect on price. (Long logs are required, and appropriately valued, for specially prestigious reconstructions, but such sales are few.) The highest prices are paid for veneer logs (say £400 per m^3), most of which are exported to Continental Europe for processing there, while logs of lower quality are used for joinery (up to £150 per m^3); lower grade still are taken for fencing or mining timber (£40), and the remainder for firewood (£20). This emphasises the value of:

- aiming management at the highest quality possible,
- knowing the market or hiring the marketing skill of someone who does, and
- taking competitive tenders for hardwood logs of all but poor quality.

There is a wide difference between the marketing of most conifer timber and that of broadleaves. Conifer logs are, in the main, standard products which, if they are to be sold at all, must conform to the miller's specifications. Hardwood logs are more variable and there is far wider scope for judgement, for finding a specialist niche in the market and for developing local outlets. There is a wealth of hardwood material which is presently chopped into firewood because neither the grower nor the cutter knows any better or bothers to market it effectively. Log size, attractive grain, colour, absence of defects such as shakes and splits, cylindrical shape and so on are all potentially important in hardwood logs. The range of values indicated in Table 16.2 emphasises the wisdom both of aiming to grow hardwoods of high quality and of seeking the advice of an *independent* consultant if you intend to sell mature hardwoods which are potentially of value.

Veneer logs should be sold 'green', soon after felling, in order to minimise splitting and discoloration. At the mill, after soaking in water and steaming to soften the wood, logs are either rotary veneered or sliced. For normal plywood manufacture, veneers are rotary cut; the log, normally three metres long, is turned against a full-length knife which moves in about 1mm each revolution, so that a continuous sheet of veneer is unwound. Large log diameters are more economical but any split or shake causes the sheet to break, requiring guillotining and gluing. Hardwood veneers for furniture are sliced; the part-log, perhaps a quarter, is repeatedly forced past a heavy fixed knife, slicing off a thin sheet which may be less than 0.5mm for decorative figured wood. With sycamore, always cut away a piece of bark to check for the presence of ripple-grain wood; this grain pattern affects the whole tree and can increase the value many times, to £1,000 or £1,200 per m³.

Table 16.2

RELATIVE VALUES OF DIFFERENT GRADES OF HARDWOOD LOGS IN RELATION TO MINING OR PALLET GRADE.

	Oak	Ash	Beech
Veneer	10	5.3	–
1st quality butts	6	4	2.5
Beams quality	2.5	3.3	2
Fencing quality	2	1.3	1.3
Mining/pallet timber	1	1	1

Source: Venables, R.G. (1985), 'The Broadleaved Markets', in *Growing Timber for the Market*, ed. P.S. Savill, Institute of Chartered Foresters, Edinburgh

Standing sales are easy to arrange and involve the grower in the minimal costs. Inevitably the prices obtained in sales of standing trees are markedly lower than those of felled trees, not only because the buyer is faced with the costs of cutting but also because, especially with broadleaves, the quality of the inside of the stem is unknown. There is good sense, therefore, in trying to sell mature broadleaves felled, so that the buyer may inspect for rot. Potentially low-quality broadleaves may be sold standing.

INDUSTRIAL ROUNDWOOD

For many purposes where formerly solid sawnwood was used, particle-board is now used: flooring, skirting-boards, kitchen cabinets, etc.; this may be 'chipboard' or the heavier medium-density particle-board. Essentially these are made of wood particles bound together with a resinous adhesive and pressed into sheets or strips. They may have a special surface applied, either a veneer of wood or a thin sheet of plastic or metal. Similarly, although of much lower density, softboard is made of wood particles loosely bound by glue and made into sheets principally for constructing lightweight walls and ceilings.

269

A wide range of papers are made from wood pulp; again the wood is chipped or shredded, before being cooked in steam or treated with chemicals to separate the fibres and tracheids of the wood. These are then mixed with water and induced to form a mat which can be drained and dried as a sheet of paper. The major softwood pulpmills require spruce almost exclusively; a pulpmill in Wales accepts almost any species of hardwood. Whereas handmade paper has always been made by the sheet, modern paper-making by machines is a continuous process with the strip several metres wide emerging from the machine at around 15 metres a second. A tear in the paper results in enormous problems as paper continues to spew out, threatening to fill the factory like a giant waste basket. The tear strength of the paper depends crucially on the fibre length of the raw material.

Recently a new product has been manufactured in Britain. Flakeboard (or oriented strandboard) is similar to particle-board, although the units are flat flakes of wood laid like small sheets of paper and bound with artificial resins. These panels are a substitute for plywood for many purposes. In Canada, where the process was pioneered, the raw material is commonly aspen, but in Britain it is pine.

All these processes use roundwood billets supplied in lengths from 1m to 3m with a range of diameters between 2.5cm and 40cm, each manufacturer having specifications for purchased wood, dependent on the machinery. Most processes require the removal of bark at the mill. Merchants generally look for minimum purchases of 100m³. It is a misconception that, because the billets are to be chipped, any old material should be acceptable. Straightness (for passing through the debarking process) and attention to the dimension limits are important, and manufacturers require conformity to restrictions on species. The type and shape of cutters for chipping require to be matched to species and the process may be sensitive to resin content, for instance larch in paper-making. The strength of paper depends on the proportion of juvenile wood and knots in the wood, and the heavy cost of clearing the results of paper tears in this fast, continuous process makes tight specification of the raw material quite essential. These are bulk markets and it may be that several proprietors of small woods should co-operate in selling to them.

As an example of the influence of wood specifications on value, one overseas paper-maker was so concerned at the cost of juvenile wood in the reduced strength of their paper and in the machine downtime due to paper tears in processing that the firm made the huge investment of creating their own plantations. In that way they could control the silviculture and avoid having to buy in the young, fast-grown material from trees planted at wide spacing, with its abundant juvenile wood and the ultra-short fibres of large branch knots. Rubbishy raw material was, to them, costly at any price. The end-use sets the value of wood.

SPECIAL PRODUCTS

The small woodland grower, while not neglecting the regular markets mentioned above, should pay special attention to some which are now virtually out of the reach of the large forest growers. Farmers and their contractors continually need supplies of fence straining posts, fence posts and fence stakes. Local authorities need street bollards, footpath edgings, tree stakes (however over-specified in reality), plant tubs, guard rails, gates, seats, benches, paving blocks, etc. Gardeners need bean-sticks, rustic fencing, path edgings and many other items. Alder is re-emerging as a useful furniture timber, similar to mahogany, as also are gean, birch, walnut, sycamore and maple.

An important developing market for lower-quality broadleaves is for wood flooring, as strips, parquet and blocks. For this, short lengths and a wide range of species are acceptable, such as ash, beech, sycamore, maples, birch, alder, etc.

The opportunities for supplying non-commercial markets should be carefully considered. Home-grown hardwoods for amateur carving and turnery are sold in the range £900 to £1,500 per m³, and the common complaint of the craftsmen is that they have to buy the materials from firms in distant towns and pay costly carriage on top. These are not mass markets but a few sales at a high price are worth a lot of firewood. Unusual species should never be casually assigned to low-grade use.

MILLING ON SITE

There is no doubt that the grower of high-quality mature timber will be well advised to sell the logs to a specialist miller who knows how best to cut them and where to sell the produce.

For lower-grade logs, however, the matter may not be so clearly decided. There have come on the market in recent years several machines which are, in effect, fully mobile sawmills, running off electricity or petrol, hydraulic or non-hydraulic. Some are self-propelled on a truck; others can be towed by any four-wheel drive vehicle, directly to a felled tree; these are horizontally mounted bandsaws. With them, growers may saw their own logs to produce timber which can be used at home or sold on. Since the machines can saw logs up to around one metre in diameter, they can handle much of the wood which has, in the past, been left to rot or has been burned. The machine can come to the log lying in the woodland.

This type of mobile sawbench may have a capacity greater than the requirements of a single small grower, but a syndicate of farmers and other owners of small woods should consider the formation of a milling co-operative, each being the joint owner of the machine; since sawing wood is not seasonally sensitive, such an arrangement may be workable. Alternatively, the machines appear to offer an opportunity for a small operator to provide a service within a district, to saw logs on contract, leaving the material in the ownership of the grower for the repair of his property or undertaking to sell on timber not required at home to a merchant for a percentage fee. There is now little excuse for the sight, so long an affront to foresters and all concerned with sustainable management, of excellent timber which has taken a century or more to grow going to waste. Forest industry journals carry advertisements of the choice of these mobile sawmill machines.

In the southern half of England there is a large concentration of woodlands which lost much of their marketable value at the end of the nineteenth century and which have had little maintenance since. Many are coppices or coppice-with-standards; few now contain stems of high timber value. From individual

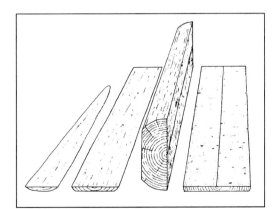

PATTERN OF GRAIN
The pattern of grain in boards cut from this hardwood log depends on how the boards are cut. The board on the right is quartersawn, those on the left plain sawn.

woods it has been difficult or impossible to sell the produce or even to justify the cost of searching for markets, but collectively these broadleaved woodlands form an important productive resource.

Very recently there has been a welcome change in this picture. As a result of collective action, led by local woodland organisations and bodies such as the Coppice Association, markets have been reopened for the produce: brushwood mattresses for flood-prevention engineering, chips for fuel and for particle-board, firewood, charcoal and so on. Britain imports more than 50,000 tonnes of charcoal each year, 95 per cent of consumption, mostly for barbecues and mostly from the unrenewable exploitation of tropical forests; home woodlands could supply all of that demand by the renewable management of coppice woodlands. Modern charcoal-making is done in steel chambers, producing a high-quality product; it is an excellent subject for collaborative action among owners of coppice woods.

By co-ordinating the management of woodlands, economies of scale can be achieved for the marketing of bulk products from the multitude of small woodlands, giving an assurance of a continuity of supply to customers which the individual small grower could never achieve. High-quality logs almost sell themselves, buyers advertising for them and ringing your bell; low-value produce needs to be sold actively.

The small woodland forester should beware of the periodic statements on the fluctuating prospects for selling to the national bulk markets. The local or regional markets for produce are much more important and more in the command of the local growers. There is a strong incentive to process wood close to the stands, in order to minimise transport costs. It is greatly in the interest of small woodland management that local markets and local wood processors be encouraged.

When a farmer sends stock to market he knows how many animals there are, he knows what they are – gimmers, ewe hoggs, cast ewes – and he knows what they *should* fetch at sale (enough to be genuinely pleased or disappointed at the close of the auction). In the same way, whatever the end-use, it is essential that the seller of trees should know what he has for sale, including the number, the approximate volume, the species and a good notion of the quality. With that information, the goods for sale should be clearly presented and the tree-grower will learn what the customer wants, what will sell and what factors affect prices.

One major benefit from growing fine wood comes from the merchant's cheque after selling it for a good price, but some growers gain even greater satisfaction from seeing wood from their own trees in daily use. An elderly farmer friend never tired of showing visitors the roof timbers on his farm buildings and slapping the handsome sliding doors on the tractor shed: 'All grown on the farm, and good for a hundred years' – and that was probably an underestimate.

FURTHER READING

Edlin, H.L. (1949), *Woodland Crafts in Britain*, Batsford, London

Forestry and British Timber (monthly journal), Miller Freeman Publishers Ltd, Tonbridge, Kent

Forestry Commission (1992), *Classification and Presentation of Softwood Sawlogs*, Forestry Commission Field Book 9, Forestry Commission, Edinburgh

Kerr, G. and Evans, J. (1993), *Growing Broadleaves for Timber*, Forestry Commission Handbook 9, HMSO, London

Savill, P.S. (1985), *Growing Timber for the Market*, Occasional Paper, Institute of Chartered Foresters, Edinburgh

Harvesting   17

Trees have to be cut for many reasons: normal harvest at maturity, thinning for the benefit of the remaining stand, the removal of dangerous stems, and so on. Above all, trees are cut to allow space for a new generation, as part of sustained yield management; if cutting is stopped, renewal of the resource is prevented until the trees fall down in senility.

Whenever trees are cut it is vital that the work is done safely and with minimal damage to the woodland and the wider environment. For tree-cutting, professionals buy chainsaws, and many amateurs do the same, sometimes with dangerously inadequate preparation. If you buy a chainsaw, you *must* also have all the protective clothing and wear it whenever you use the saw. Furthermore, make sure you get proper training before you use the machine. Information about courses can be obtained from the Forestry and Arboricultural Safety and Training Council or from a local agricultural or forestry college. A chainsaw is a powerful tool which deserves high respect; never use it casually and never use one without training.

It is important to know that there is a clear obligation on landowners and householders to ensure that people who work on their property take proper care when carrying out maintenance, cutting trees and so on. Specifically, under the Health and Safety at Work Act 1974, contractors and employees using chainsaws must wear protective clothing and follow safe practice; the standards are set out in the Personal Protective Equipment Regulations 1992.

Protective clothing for chainsaw operation should comprise:

- safety helmet with ear covers and eye protective visor (BS 5240)
- clothing incorporating loosely woven long nylon fibres for shoulders, neck, arms and upper chest
- leg protection made of the same nylon material, either as 'chaps' on the front of the leg or as full trousers, the latter giving the better protection required by tree surgeons and all occasional users. If it is cut, safety clothing of woven nylon blossoms out so quickly into thousands of fibres that the chainsaw is jammed instantly.
- chainsaw boots giving protection to the toes, top of the foot and front of the lower leg (alternatively, protective gaiters worn with steel toe-capped safety boots for occasional users)
- gloves with a protective pad on the back of the left hand

Chainsaw accidents are more likely to occur with a saw which is blunt or poorly serviced, because jams cause the kickbacks which can jerk the saw back uncontrollably to strike the operator's head or shoulder, so attend properly to maintenance and sharpening. Accidents are especially likely when the sawyer has unsteady footing. Unless you are a professional and have heavy insurance, never use a chainsaw while up a tree.

Since there is no guarantee that every reader is qualified to use a chainsaw, this chapter is written on the assumption that none is available; we shall work with hand tools. Those who can use a chainsaw can easily interpret for their own purposes.

Harvesting is not an end, it is part of the cycle of production. Just as the condition of the stand created by decades of past tending determines what harvesting is due and how it may be done, so the harvesting operation deeply affects the operations which follow it to establish the new stand. A cheap and nasty harvesting operation which leaves woodland in a mess – high stumps, uncollected logs, lop and top anyway and the land surface like a tank training area – means high costs and endless bother in regeneration.

Before harvest begins, think through the following:

- What trees are to be cut and what will they produce?
- What markets will the produce go to?
- Into what sizes will trees be cut?
- How will timber be extracted from the wood and how will it be transported?
- Should produce be sold standing?
- If trees are to be felled 'in-house', what tools will be required for the job?

The owner must decide whether the timber is to be worked in-house or contracted out. It may be possible to contract the cutting while the produce (in all or in part) is retained. More usually the choices are to sell the trees standing, or to cut them by one's own labour and sell logs at stump or rideside, or to cut and work the wood into products with one's own labour and sell the prepared products.

If you have no experience of cutting trees, do not try to acquire expertise by first working with large trees; start with early thinnings and coppice. Where work is to be done by an outside firm, a formal contract is desirable in order to avoid misunderstandings. It may be let by negotiation (useful for a small job with a trusted contractor), by simple tender (where the fair price is well known and one contractor is asked to submit a written quotation, perhaps a firm already working with you) or by competitive tender, where several firms are invited to tender in writing for defined work.

If harvesting of any kind is to be done by outside merchants, it is essential to have a proper legal sales contract which covers the following points, the detail depending on the size of sale:

1. Who are the parties to the agreement, and on what date is it made?
2. What is being sold and bought: standing trees, cut logs, etc.?
3. The location, name of the property, referring probably to a map attached, with the woodland clearly marked.
4. The species, the number of trees (or logs or poles) and the volume, how the trees are to be identified (colour paint spots, blazes, etc.). This should be backed by a clear statement that the purchaser has in fact (or is deemed to have) satisfied himself as to the accuracy of these figures.
5. The price to be paid and when; it may be that part of the price will be paid at the starting date of the sale and the remainder in timed instalments, if it is a large contract, or at completion.
6. In a sale of pulpwood or similar produce, the basis of payment may be the weight of the material as determined by passing the truck over a weighbridge at the mill gate, backed by weighbridge dockets.
7. The route by which timber is to be removed and any restriction on weights, dates and type of vehicles; for instance, there may be a culvert capable of carrying

only 15 tonnes, or from a farm woodland the timber may have to be taken over stubble only after harvesting the field crop, making the timing of the timber-working critical. Liability for damage to fences, gates, roads, etc. should be clear, as should liability for damage by fires.

8. Prohibitions may be important (e.g. tractors to cross a stream only at set crossing points – and only wheeled tractors – and no dragging of timber through a stream; debris to be cleared from ditches and streams, etc.). The penalties attached to the buyer cutting trees other than those marked for sale should be stated (e.g. payment at independent valuation and the trees remaining the property of the seller). It is usual to restrict the siting of debarking and other machines and to specify the dispersal of waste.

9. The purchaser should agree to comply fully with the Health and Safety at Work Act and to indemnify the seller against claims arising from the purchaser's employees in working this timber. He should also agree to take proper fire precautions, to employ no known poachers, not to bring dogs on the site and to have an approved insurance policy to cover negligence.

10. Finally, the contract should state starting and completion dates for working the timber which is sold, and state to whom the timber that is not removed by the completion date belongs.

This may seem elaborate and it may not all be necessary for a small sale, especially if the buyer and seller have a long-standing business acquaintance. On the other hand, the sale and working of standing timber is quite different from selling livestock or cereals; the timber may have to be worked over several weeks and the buyer's employees will be coming and going (or not going!) over that period. The risk of incidental damage or conflict of interests is far greater than in the sale of other farm produce. Some merchants have been known to treat a purchase as a standing timber reserve, to be worked at their convenience, so it is entirely reasonable to fix a date when the operation is to be complete, after which any material not removed reverts to the seller.

Sales of cut produce sold at stump or at rideside may be covered by a similar contract or agreement, except that measurements may be made mutually and agreed on the spot, so that only the price per cubic metre has to be agreed for each species.

TOOLS

Tools regularly required for tree harvesting are as follows:

Bowsaw: a tubular steel bow with a detachable blade, used for felling and cross-cutting stems between 50mm and 300mm or so in diameter. Sizes range from about 600mm to 900mm in length (the size of stem they can cut is limited by the depth of the bow); 750mm is suitable for one-person working, the 910mm is best for two people. Blades which can be sharpened are available, but it is easier to fit blades with tungsten-tipped teeth which may not be resharpened.

A *triangular bow saw* with a 530mm blade, also detachable and not able to be sharpened, is shaped suitably for working coppice.

A *two-man cross-cut saw* is essential for larger trees if a chainsaw is not available. The length should be chosen according to the size of trees to be cut; 600mm longer than the largest tree to be cut is adequate, so 1.2m or 1.3m long is sufficient for most work. Too long a saw for the job is cumbersome. Handles are attached to the saw by bolts and wing-nuts, so that the blade can be slipped from a cut if the tree 'sits back'. Blades may be straight-backed or concave; concave is

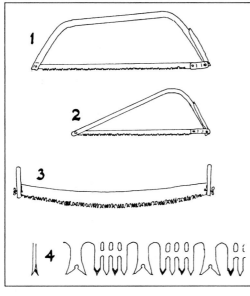

SAWS
1. *Bowsaw, up to 910mm long.*
2. *Bowsaw about 530mm, suitable for cutting coppice.*
3. *Two-man cross-cut saw, hollow-backed with detachable handles.*
4. *Raker design teeth: the large spaces on either side of the raker teeth are to accommodate the ribbons and crumbs of cut wood. Rakers must be marginally shorter than the peg-teeth. Left: the set of the peg-teeth.*

preferable because friction is reduced and the design interferes less with wedges put into the cut. There are two main designs of teeth, peg-teeth and racer (or raker); the former is traditional for hardwoods and the latter more effective for softwoods.

For maintaining the cross-cut saw you require fine-cut triangular saw files for sharpening, saw-setting pliers and a home-made saw vice – or an arrangement with a local saw-doctor.

Wedges: aluminium alloy or hard plastic (to avoid accidents if a chainsaw should touch one; formerly steel wedges were used.) Wedges are between 75mm and 200mm long and require a heavy hammer for driving them in.

Axes: These are scarcely, if ever, used in current forestry, since all their work is now done with a chainsaw. Modern Scandinavian design axes are recommended in place of the traditional heavy English pattern. A 2kg head is adequate for felling and a 1.5kg head for snedding (cutting off side branches from the felled tree). Most people find a haft of 760mm convenient for a felling axe and 675mm for snedding; all axes should have fawn's foot ends to the hafts, to prevent the swinging axe slipping out of your hand. Traditional English felling axes had narrow heads weighing between 2.5kg and 4kg on long, straight hafts; designed for cutting hard oak, they are cumbersome and exceedingly tiring to use by non-professionals who can achieve greater accuracy with a light axe-head in a shorter haft. Raining blows roughly on the target wastes energy and wood, and is usually the result of tiredness.

Whetstone: a canoe-shaped whetstone will be needed for sharpening axes or a traditional circular carborundum stone. Also required

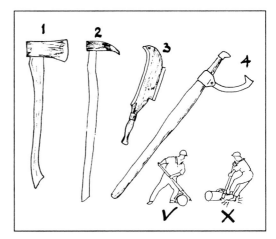

HAND TOOLS
1. *Scandinavian style axe: short haft with fawn's foot.*
2. *Sappie for moving light logs.*
3. *Billhook for cutting light coppice and trimming poles.*
4. *Peavey: below, the right and wrong ways of using it.*

will be the use of a slow-speed grindstone.

A *billhook* is required for cutting light scrub, small coppice and hedging and for marking blazes on trees for thinning. The design is a matter of personal preference and regional tradition. These too are sharpened on the grindstone and with the canoe-shaped whetstone.

A *sappie* is a short, heavy spike on the end of a stout handle, for moving light logs. For trees and heavier logs use a peavey, a curved metal arm with a spike, hinged to a heavy pole.

Tongs are sold for carrying logs; their use is a matter of preference. They are easily mislaid in harvesting debris; for safety, care must be taken that they grip the log securely, lest it falls on your foot. They tend to be a fiddle for light logs but encourage people to attempt carrying logs that are too heavy, which is a recipe for back injury.

Tapes are needed for measuring timber; you will require both a long metric tape for the length of logs and a diameter tape (see chapter 15). The professional woodcutter uses a spring-loaded tape attached to his belt, the zero having a hook to fasten to the end of the log so that the tape reels out as he works up the stem; these are expensive and unnecessary for small-scale operations.

Peeling spade: this is for removing the bark from certain types of produce; the Swedish pattern with detachable blades is best for most purposes, although the heavier English design may be better for stripping oak tan bark, a process which involves levering the bark off rather than peeling narrow strips.

A *first-aid kit* should be available in the woods when cutting tools are being used. Wear a hard hat when working timber, even if you are not using a chainsaw.

TOOL CARE
Blunt tools are dangerous because the worker must try to cut by force instead of getting the tool to do the work. The first rule for all wood tools is that the cutting edges must never be allowed to touch anything harder than wood. Always clear away litter from around the base of the tree before beginning to cut, lest there is a hidden stone. Never drop an axe on the ground: there will be a stone among the leaves and it will knock a 'bite' out of the edge – stick the axe blade into a stump.

When necessary, grind back the 'shoulders' of the axe blade, behind the cutting edge, on a grindstone, using plenty of water to keep the steel cool and the stone clean. (A high-speed stone will destroy the temper.) Then set the cutting edge with the whetstone.

If not mis-used, hard-tipped bow saw blades do a lot of cutting before requiring replacement. No maintenance is involved.

The cross-cut saw blade should be cleared of resin using paraffin and white spirit on cotton waste. The teeth require to be regularly sharpened and checked for the correct setting, but if used with care (kept clear of soil etc.) the saw should not require sharpening or setting more than once a week even in moderately heavy use. Full sharpening can be arranged with a professional saw-doctor but maintenance with a file and setting pliers is well within the capacity of anyone doing forestry work.

The set of the saw teeth is as important as sharpness. Looking along the length of the saw, the rows of teeth should form two exactly straight lines. A tooth with less set than the rest will do no work; a tooth with too much set will be out of line and force the sawyers to do extra work, or even jam the saw. If one row of teeth is set more than the other (or if several teeth are over-set), the saw will try to cut round a curve and will continually bind. The raker teeth work as chisels, set slightly shorter than the peg-teeth, to rake out the cut wood; properly set, they bring out crumbs and

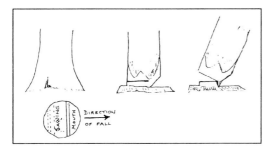

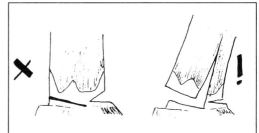

LAYING-IN AND FELLING

Centre: The saw-cut must be above the bottom of the mouth. The tree should then split down from the saw-cut to the mouth.

Bottom: Keep the saw-cut parallel with the mouth.

FELLING: WHAT TO AVOID

If the saw-cut goes below the bottom of the mouth, the tree may well split upwards, ruining the butt log.

'worms' of wood, whereas a blunt saw produces dust.

Axes and billhooks require sharpening in the field much more frequently than saws. Several times a day the edge should be 'touched up' with the whetstone. Judge sharpness by looking at the edge against the light; if it reflects a line of light, it is blunt.

It is wasteful to cut down trees with an axe alone; use a saw, but (in the absence of a chainsaw) the axe should be used to cut away the swollen root buttresses (which would increase the thickness to be sawn and which add nothing to the useable volume of the tree). Before swinging an axe, make certain there is no obstruction in the way, no branch and no person.

Never begin to cut a tree within falling distance of telephone or power lines without informing the company and having their formal agreement. Never begin to cut a tree while there are people nearer than two tree heights. Before beginning to cut a tree, clear debris from around it and from your escape route, away from the planned line of fall.

After the buttresses have been trimmed to just below the sawing level, the tree should be 'laid in' to make a 'face' (or a 'mouth' in other districts) in order to induce the tree to fall in

the desired direction. A tree cannot be made to fall in a particular direction if it leans strongly in another or is heavily weighted by its branches. If it must be made to fall against its inclination, the tree must be pulled by rope over a pulley block while it is being sawn. Nevertheless, a reasonably balanced tree can be induced to go in a given direction by good preparation, by laying in and careful sawing.

The face should be low, horizontal and straight, at right angles to the intended falling direction, cutting into the stem to a depth about one-fifth of the basal diameter. The sawing should begin exactly opposite the face and at a level which will bring the saw-cut about 3cm above the bottom of the face for medium-sized trees, 5cm above for larger trees. If the tree shows a tendency to sit back on the saw or a reluctance to go in the desired direction, wedges may be put in the saw-cut and tapped in behind the saw while sawing continues. Do not force the wedge, otherwise the tree may split up the stem.

The saw-cut must not wander below the bottom of the face; the result of that is usually that the butt splits upwards, ruining the most valuable timber. If the work is well done, the tree should fall by splitting down to the mouth, turning on the straight hinge of the

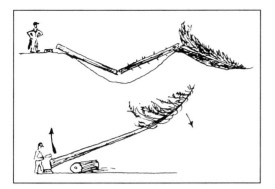

OBSTACLES TO FELLING
Felling across a dip may cause the stem to snap. Never fell a tree so that it falls across a log or a rock: the butt end may fly up – a safety hazard.

face which has been cut to determine the direction.

When using a cross-cut saw the sawyers should bend or kneel close to the tree, with their backs towards the direction in which it is to fall. Each sawyer must pull only, never pushing. The secret is to swing the saw rather than pull it (never jerk it). When the tree begins to fall, which takes several seconds, the sawyers stand up and by taking a few paces forward are in the safest place to escape from danger, behind the stump and slightly to one side.

Avoid cutting a tree so that it falls across an obstacle such as a stump, rock or log, because

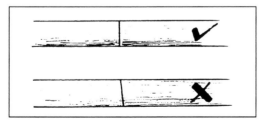

CROSS-CUTTING
When cutting the tree into logs, be sure to cut at exactly right angles to the axis of the stem. Cutting at an angle (below) wastes timber and may cause the customer to reject the log altogether if it is short.

the butt is liable to jerk violently upwards. If the saw jams in the cut and the tree is falling, leave the saw where it is; the worst that may happen is that some teeth may be bent. Avoid cutting a tree so that it falls across a depression in the ground; the stem may snap across at the dip.

Wind makes tree-felling difficult and, if it is strong, dangerous. Do not fell trees on top of one another; clear each one before felling the next.

In group felling or thinning, difficulty may arise if a tree becomes 'hung-up' on neighbours. If it is small, it is usually possible, having made sure that the tree is completely cut free of the stump, to drag the butt directly away from the point of hang-up and thereby induce the tree to fall. With larger trees this process becomes more difficult and hazardous; a middle-sized tree may be brought down by pulling back the butt with a monkey-winch, but much depends on how deeply the butt digs into the soil. With a large tree there may be no safe alternative to bringing in a heavy tractor and winch. Even with a small tree, never try to solve the problem by cutting the tree on which the first is hung up; this is seriously dangerous.

Except for large branches on hardwoods, trimming the branches on the fallen tree – 'snedding' – should be done with an axe (if you are not using a chainsaw). Always work from the butt up the bole, cutting branches on the side further from where you are standing and away from your legs, *never* towards your feet and legs. Cut branches at the surface of the wood (not level with the surface of the bark). Before you cut into any branch, consider whether it is supporting the tree or whether the tree will roll when that branch is cut.

Trees on sloping ground should, if possible, be felled uphill or, failing this, along the slope. Sawyers must take particular care to have a clear escape route, since the tree may slide downhill or roll after hitting the ground.

Windblown trees present special dangers; a tangle of blown and half-blown trees should never be tackled by half-skilled people. Even a single stem may be difficult; in an unsupported leaning stem the top side will be in tension and the bottom in compression, but if the upper bole is supported the reverse will be the case, binding the saw blade. A windblown stem when cut through may jump or roll and the root-plate may shut like a five-tonne rat-trap. Harvesting such timber calls for great skill.

SEASON FOR HARVESTING

All broadleaves are best cut in winter, any time after leaf-fall and before the sap rises in spring. Conifers may be cut in summer or winter, although pines cut in summer have a greater tendency for the wood to discolour with blue-stain fungus.

Several species benefit from prompt removal from the forest to the sawmill, for instance beech, ash, sycamore and cherry, and pine among the conifers. Oak and sweet chestnut both have high tannin content to resist fungal and insect attack and they are less susceptible.

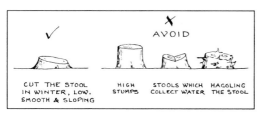

COPPICE CUTTING

COPPICE

Coppice is best cut when it is leafless. Cutting may be done with a billhook, axe or bowsaw, according to the size of the stems. Small shoots may be cut easily with the billhook, using an upward stroke, but the triangular bowsaw is most general. The stems restrict access to their neighbours, but they should be cut close to the stool, avoiding tall snags.

It is traditional to cut coppice with a clean sloping cut to shed water (the idea being to reduce decay of the stool). It is more important to avoid 'haggling' the cut and damage to the stool, since the production of new shoots depends on the cambium of the stool and on the root system below it. Shoots will regrow if cutting is done in summer but it is better to do the work in winter, so that the new shoots will be able to make a full season's growth; cutting in early spring may leave the tender shoots open to late-spring frosts, and late-summer cutting may mean the shoots do not harden before winter.

Traditional coppice production may amount to a sustainable 2–5 tonnes of dry wood per ha per annum, measured to 5cm diameter. When freshly cut the stems will have a moisture content of 100 per cent, measured against the dry-wood weight, so green-weight production is about double the dry. (For short-rotation coppice see chapter 14.)

EXTRACTION

While very small stems may be moved by hand, most logs require a horse or tractor to move them to roadside and transport beyond.

Within small woodlands most extraction will be by ground-skidding, that is to say the logs or tree stems are pulled along the ground, not carried in a vehicle; the energy required for this is much reduced if the forward end, generally the butt, can be raised above ground level, so that only the farther end is dragging and causing friction.

The use of horses for timber extraction has recently re-emerged to a limited extent after several decades of almost total disuse. The

TRACTOR SKIDDING

With small modifications, a farm tractor can be used for hauling light timber.

Top: With a hydraulic lift, the ends of poles are kept clear of the ground.

Middle: A skidding pan keeps the pole ends off the ground and reduces friction.

Below: A winch is invaluable: each log is secured by a choker chain, the end of which slots into a keyhole slider on the winch wire, so that several small logs can be hauled together.

main problem with using horses for forestry work is not their limited power but the difficulty of finding willing horse-handlers. The most suitable type of horse, in the view of most experts, is the stocky work pony, such as the Highland garron, rather than a heavy draught horse. Horses are financially competitive with tractors for extracting pole material, especially in thinnings and group management; they have the great merit that they learn, which tractors do not, and they tend to do less damage to the ground and trees. In effect they become experts at the job. Guidance on planning operations with horses and on harness design has appeared in the journal *Small-Scale Forestry*, published in Sweden where horses are used extensively in farm woodlands. Both in Scandinavia and in Britain there have been improvements in the design of harness and attachments for moving timber, including wheeled timber arches to lift the butts clear of the ground, thereby reducing drag, so that the horse can manage heavy loads more easily. (See p.306 for address of British Horse Loggers Specialist Group.)

Light farm tractors can be used very effectively for timber extraction in woodlands. Necessary design features are:

- a front guard to prevent heavy sticks puncturing the radiator
- a sump protector, a guard plate welded under the engine to take the shock of striking a stump
- a winch, preferably with some device to lift the wire and enable the butt ends of poles to be clear of the ground, thus greatly reducing the drag. Choker chains, which run loose on the winch wire, allow several poles to be drawn into a bunch and pulled behind the tractor as a single load.

Several tractor manufacturers in Europe produce mini-skidders, especially designed for

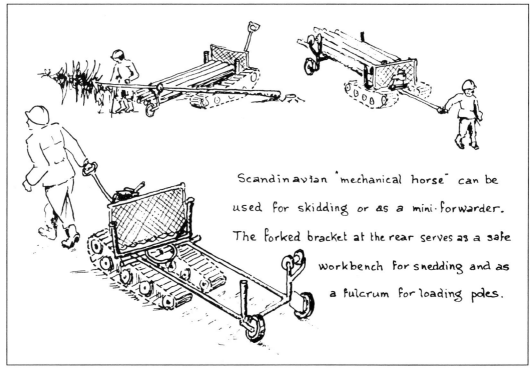

Scandinavian 'mechanical horse' can be used for skidding or as a mini-forwarder. The forked bracket at the rear serves as a safe workbench for snedding and as a fulcrum for loading poles.

MINI-TRACTORS
There are several designs of these 'mechanical horses' which can be used either as skidders or as mini-forwarders.

small woodland work. Although their performance may appear to compare unfavourably with large forest tractors in terms of their work output per hour, these figures are irrelevant for small woodlands. The powerful forest tractors are hopelessly uneconomic if they are engaged in moving small poles and if they are worked intermittently. A mini-skidder tractor, comparable in size and power to one or two good horses, may be precisely what is required for some small woodlands and are worth serious consideration. A light farm tractor, with guard plates and winch added, can do excellent work. The real message is to employ an appropriate tool for the particular task and not to be over-capitalised.

In large-scale forestry much use is now made of the *forwarder*, a tractor which carries its timber load entirely clear of the ground in an integral rear bunk or articulated trailer, and which incorporates an integral loading crane. Although these may be built around a farm tractor (certainly the case in Scandinavia), modern forwarders are purpose-built, frame-steering machines with powerful hydraulic cranes and capable of carrying loads of 10 tonnes; their ability to work off-road and collect timber at stump is particularly useful for harvesting large volumes of small timber. Their cross-country capacity and the fact the load is carried on large rubber-tyred wheels means that a contractor with one can extract timber over farm fields with minimal damage.

Timber extraction requires careful planning. The horse or machine must be able to reach the felled timber and to skid (or carry) it out of the stand to a ride or road.

This generally requires 'racks' to be cut, i.e. a row of trees to be removed, at intervals of about 25m in young conifer crops. There must be sufficient space to stack the timber (and perhaps to work it in some systems) at the side of the ride or road, in preparation for its removal by truck. The size of the stacking space depends on the volume of wood to be removed; there must be adequate space whether the work is to be done in-house or by contract.

In the past most felled timber was extracted from the woodland in tree lengths (or perhaps in half-lengths) to roadside, where it was cross-cut into logs and shorter pieces, sorted into categories and collected in road trucks. The disadvantages of this system are the high risk of the long lengths scraping the standing trees as they are skidded around a corner within the stand, and the difficulty of keeping the felling, extraction and cross-cutting in step with each other. Damage to standing trees takes the form of bark stripped from the lower bole, allowing disease to enter and seriously reducing the future value of the butt logs, the most valuable part. It is an advantage of the system, however, that the decisions on cross-cutting the stems can be made by one skilled person at the roadside depot, which is likely to provide consistency and accuracy. Inevitably some waste wood is extracted to roadside.

Commonly in extensive conifer forestry, harvesting is now done on a 'short-wood system'. Delimbing and cross-cutting are done at stump by the chainsaw-worker who has felled the tree. He works at his own speed, leaving the log lengths where they have been cut and gathering the short pulpwood lengths into small stacks of four or five. The forwarder then follows, picking up material for homogeneous loads, all pulpwood or all one log category. Only useful material is extracted; unsaleable wood is left in the stand.

Whether extraction is by forwarder or by skidding, and whether in a large forest or a small one, it is helpful if the vehicle can make circular trips, to avoid the wasted space and time of turning. In woodland being managed for continuous cover (group selection, for instance), the network of the main extraction routes should be allowed to persist from one felling to the next. This reduces the overall damage to young groups, avoiding the waste of expenditure on trees which would then be destroyed in subsequent timber working.

Pulpwood is usually sold by weight, and the lengths rapidly lose weight as the sapwood dries. It pays handsomely, therefore, to organise the harvesting, sale and transport to the mill so that the wood reaches the mill fresh, green and heavy.

Extensive forest management is now experimenting with two other timber-harvesting methods: whole-tree extraction and chipping (neither is likely to be used in small woodlands). In the former, the whole tree is skidded from stump to depot, where the logs (and perhaps pulpwood) are cross-cut and the remaining waste lengths and branches are chipped for transport to a mill to make pulp or boards. Alternatively the whole tree may be chipped, at stump or at roadside. These systems, the present extremes of harvesting, are of questionable sustainability since the twigs and leaves (needles) are removed from the forest, thereby depleting its nutrient capital. This bears on the management philosophy of the owner and national forestry authority, on whether the tree crop is harvested and it is then to be considered what can be done with the bare land (perhaps now permanently poorer), or whether the capital resource comprises the land, trees and other organisms from which a harvest can be taken periodically but which will never be degraded.

The harvesting of large hardwood trees

requires special attention. Irrespective of quality, large trees present problems in felling and delimbing; they are very heavy – bole and branches may be 25 or 30 tonnes – and there is always a serious risk that the removal of one large limb after felling may cause the whole tree to roll over. The boles of trees of medium to high quality should not be cross-cut until seen by a potential specialist buyer who can discuss or decide how best they may be divided.

In Britain most wood has to be transported rather long distances, to large pulpmills or specialist hardwood sawmills. Since it is costly to load and unload bulky material such as wood, it is most profitable in large forest units to build roads to a standard so that the 38-tonne road truck has access deep into the woodland; this includes wide gateways, solid roadway, strong bridges, sufficient width and provision for turning. The manager of small woodlands will probably decide it is more economical to accept higher costs of extracting from stump for a longer distance to a depot on or close to a public road, thus avoiding the capital cost of constructing a high-grade forest road to each small woodland. This decision bears on how timber is to be worked and on the selection of tractor and associated equipment.

Many people who have seen large forest areas after harvesting express wonder at the quantity of wood left as forest waste, much of it potentially useful as firewood. In some European countries this is sold retail to people who glean an area or pay by the load. Farmers in Scandinavia normally harvest the waste wood from their own lands to put through a chipper run on the power-take-off of the farm tractor; the chips, when dry, are used in automatic-feed stoves for all heating on the farms. The amount of wood residues after harvesting bears closely on forest hygiene (for example on the populations of pine weevil and pine shoot borer, see chapter eight) and also on the conservation of insects in general. The abandonment of waste logs after felling with the intention of benefiting wildlife may come with a large price-tag in the form of pest damage, and there may be a conservation penalty if a serious pest outbreak requires the use of insecticides. There is a balance to be struck; waste coniferous logs from which ribbons of bark have been stripped are less useful for pine-weevil breeding but will be available for non-pest species later.

All too often the harvesting of timber involves ground-skidding big timber, tractors churning the soil and heavy trucks rutting earth roads, all creating mud. Such soil erosion washed into streams would inflict serious environmental damage which can be avoided. For large forest harvesting, silt traps must be built to prevent any mud entering streams. Everywhere discipline is needed to prevent tractor-skidding through or near streams, which leads to bank erosion, and extraction may have to be stopped in very wet periods or when roads are especially soft, for instance after frost. Soil erosion is most serious where felling coupes are large and least serious where woodland is managed on a group-selection or other 'continuous cover' system. Harvesting should not involve non-sustainable damage; that is contrary to sustention.

FURTHER READING:

Dutch, J. (1995), *The Effect of Whole-Tree Harvesting on Early Growth of Sitka Spruce on an Upland Restocking Site*, Research Information Note 261, Forestry Commission, Edinburgh

Forestry and Arboriculture Safety and Training Council, Numerous safety guides on all types of woodland operation and machinery use, F and AS and TC, 231 Corstorphine Rd, Edinburgh EH12 7AT

Forestry and British Timber (monthly journal), Miller Freeman Publishers Ltd, Tonbridge, Kent

Forestry Commission (1984), *The Farm Tractor in the Forest*, Forestry Commission, Edinburgh

Forestry Commission (1994), *The Use of Horses for Timber Extraction*, Forestry Commission Technical Development Branch, Ae Village, Dumfries

Harmer, R. (1995), *Management of Coppice Stools*, Research Information Note 259, Forestry Commission, Edinburgh

Hibberd, B.G. (1991), *Forestry Practice*, Forestry Commission Handbook 6, HMSO, London

Small-Scale Forestry. Newsletter (quarterly) from the Department of Forest Extension at the University of Agricultural Sciences, Sweden (published in English)

Legislation in Britain concerning trees and woodlands has been heavily affected by three facts:

- In both Great Britain and Ireland the woodland resources are unusually scarce compared to other countries in Europe and, indeed, worldwide.
- Trees grow relatively slowly and there is always a high risk that their owner at any time (in effect the temporary steward of the woodland) either by accident or design may cut devastatingly into the woodland capital.
- Allied to the previous point, all woodlands, but especially those managed positively for multiple objectives, exert influences beyond their boundaries and offer benefits to a society wider than their owners in law.

As a consequence of these considerations, forestry legislation generally seeks to restrict the actions of owners, to prevent them playing fast and loose with their woodland property. In some instances, ownership rights are directly restricted – for example, limiting the right to cut immature trees, to fell woodland unless there is a promise to regenerate it or to cut trees covered by a Tree Preservation Order – whereas in others the Forestry Authority seeks to have its way by offering grants for those who accept the public weal. Most Forestry Acts have sought to protect woodland from casual destruction and to promote new afforestation and the proper regeneration of existing woodland. Some legislation, such as

the Wildlife and Countryside Act 1981, has given powers to restrict the management of specifically designated areas of woodland (and of many other habitats) to ensure their continued existence for their ecological and natural conservation values.

Many statutes relating to farming also apply to forestry. For instance, the Weeds Act 1959 places a duty on all occupiers, including those of woodlands, to control noxious weeds such as spear thistle, creeping thistle, curled dock, broadleaved dock and ragwort, and under the Wildlife and Countryside Act 1981 it is an offence to allow Japanese knotweed and giant hogweed to grow wild.

In Great Britain the offer of financial advantage for forestry action, both by direct grants and, importantly, through taxation, focused for over 70 years on afforestation and regeneration. This contrasts with practice in Scandinavia, for instance, where, in very different circumstances of abundant forest wealth, owners gain most by ensuring that stands are felled and regenerated promptly when they become mature; an owner who cuts prematurely or who holds overmature woodland is penalised in severe taxes, since these actions are seen as selfish and anti-social in countries where the gross national product depends heavily on a sustained flow of mature trees to the wood-processing industries.

TAXATION

Here is surely the clinching attraction of woodlands: in the United Kingdom they are not subject to income tax! This extraordinary situation arose from the changes introduced in the 1988 Finance Act.

Before 1988 the owner of woods had the choice of paying tax on a low assumed income, that assessment being sustained year after year irrespective of actual profit or loss, or of lumping the actual profit or loss of the woodlands with his other taxable income (a procedure denied to farming for many years). The first option was attractive for mature woods likely to make a large profit, while the second was ideal for very high income-tax-payers wanting to invest in new afforestation, since the investment costs ('losses' in the current tax year) could be offset against other income.

Large-scale forestry planting, mostly in the uplands, was extremely advantageous to the highest-band tax-payers; at one time the highest-band tax rate was 95 per cent, so the offsetting of afforestation costs meant that £100-worth of forest could be acquired by that tax-payer for £5 net. The plantations could then be sold at their full value some years later without incurring further tax. Much excellent forest, and some poor, was created as a result.

The financial attractiveness and success of the latter arrangement was its own undoing. The speed of the environmental change caused by afforestation was disturbing in some places and political lobbying led to the Budget decisions in 1988 which virtually took forestry and all woodlands out of the tax system entirely. The costs of planting woodland, either as new afforestation or replanting, are no longer allowed as a deduction in income-tax assessment and the income derived from sales of wood is not taxed either. Also not taxed are the Forestry Commission's planting grants and woodland management grants, which were substantially increased after 1988 in order to restore the

incentive to plant following the withdrawal of the tax advantage, particularly to encourage farm forestry and afforestation 'down the hill' rather than in the uplands.

For an exclusively woodland enterprise the situation is clear: there are no income-tax allowances and there is no income tax on forestry income; production forestry is a taxation void. For woodland work on a farm, however, where farming costs are offset against farming income, the situation may be less clear-cut and some farm-woodland costs may be carried effectively on the farm account. Woodlands Grant Scheme income received from the Forestry Authority is not taxable (being counted as woodland income), but grants paid by the Agricultural Departments (MAFF, SOAFD, WOAD), such as the Farm Woodland Premium Scheme and woodland set-aside options, are regarded as farming income and are taxable. Income from game-shooting in woodland is taxable. Income from Christmas trees cut from woodland managed principally for timber will probably be treated by the Inland Revenue as tax-free, since it is forest income, as distinct from trees produced in a Christmas-tree farm.

The value of standing trees and woods is not liable for capital gains tax, and never has been. Realised gains on the land, after indexation, are subject to capital gains tax, but the land value is usually a small proportion of the total value of maturing woodland. In respect of inheritance tax, woodland capital is treated in a similar fashion to farming capital but with some advantage in favour of woodlands; the differences are sufficient to make it advisable for a farmer contemplating a substantial investment in new woodlands to consult a tax adviser in order to get maximum advantage.

In brief, when a woodland investment has been held for two years or more, its value qualifies for 100 per cent business relief from inheritance tax, and if the woodland is made over as a gift after that qualifying two-year period, the business relief is passed to the recipient. (The normal rule that the donor must survive for seven years from the date of an absolute gift in order to qualify for inheritance-tax relief does not apply.) A woodland owner enjoys tax-free income during his lifetime and the capital value remains protected from inheritance tax. Taken with the exemption from capital gains tax on the timber value, these are considerable attractions.

SAFETY

Work in the woods should be healthy and safe, but several operations, especially those involving power tools and chemicals, are potentially dangerous. It is common sense, as well as a legal responsibility, to take proper care.

The Health and Safety at Work regulations make it clear that the occupier of the land, as well as an employer, is responsible for ensuring that safe working is practised, whether the people working are employees, volunteers or sub-contractors. If unsafe work is being done on your land, it is up to you to stop it. Equally, there is a duty on individual workers to ensure they work safely; it is no excuse in law to say that you knew it was unsafe but you were told to do it.

Special care should be taken with chainsaws, tractor power-take-offs, tree-felling and piled logs, and chemicals.

No one who has not received proper training on a recognised chainsaw operator's course and who is not wearing full protective gear should use a chainsaw. These are wonderful tools but chainsaw accidents are horrendous. The odd-job ('It's not worth putting on all the gear') is the most dangerous of all. If you are not trained, use a bowsaw; conceivably you may draw blood,

but you will not slice off a foot or cut your brain in two!

Several useful woodland machines – saws, soil augers, chippers, etc. – work from the power-take-off (PTO) of a farm tractor. The PTO guards must always be in place.

Trees are heavy; a tree of about 25cm diameter breast height (dbh), with its twigs and foliage, probably weighs a tonne. When falling, even a middle-sized tree has immense power and is capable of jumping about if it hits an obstacle. Never begin to fell a tree if there is someone else within twice its falling distance. 'I thought it would fall the other way' and 'I didn't think it was so tall' are excuses that will do no good and will ring in your mind for the rest of your days.

When a tree is safely on the ground and you are trimming off the branches, you must be constantly thinking about whether the branch you are about to cut is acting as a prop or whether cutting it would change the balance of the tree and cause it to roll over.

Piles of trimmed logs ready to be picked up for transport to the sawmill are an attractive sight to warm the owner's heart as the product of the woodland and of hard work, but they are also an attraction for the adventurous who want to climb and for the weary who want a seat. Log piles are usually unstable, not meant for climbing, nor as picnic seats. If the woods may be open to visitors – not only public places but places where casual visitors may come – put a notice on log piles to warn people of the danger of logs rolling – and do not climb on piled logs yourself.

There are necessarily strict rules about the use of pesticides and herbicides, and about the disposal of their 'empty' containers. Anyone working with these chemicals on someone else's property requires a certificate of competence, having been trained in their use, and if you are working on your own land you should want to know what you are doing. The Forestry and Arboriculture Safety and Training Council (see appendix) or the local agricultural or forestry college will advise you about these and other training courses concerning safety. The rules governing the use of herbicides in farming are not all the same as those in forestry. Do follow the instructions on the label and never decant these chemicals into containers other than the manufacturer's.

Apart from these safety responsibilities in respect of work, the occupier of woodland has a duty to visitors under the Occupier's Liability Act 1967; the occupier must take reasonable care to see that the visitor is reasonably safe in using the area he is permitted to use, even though he has not been invited to come.

TREE FELLING

By various Forestry Acts Parliament has given powers to the Forestry Commission, as the national authority, to prevent woodland being destroyed. These are now vested in the Forestry Authority, the non-trading part of the Forestry Commission. It controls the felling of trees throughout Great Britain and requires the owners of woodland to obtain a felling licence before work is done. There are some legal exceptions to the rule:

- where the felling is in accordance with an approved plan of operations under one of the Forestry Commission's grant schemes;
- trees in a garden, orchard or public open space;
- when the trees are all less than 8cm dbh or, in the case of thinnings, below 10cm dbh or, in the case of coppice or underwood, below 15cm dbh;
- trees which are interfering with permitted development or statutory work by public bodies;
- trees which are dead, dangerous, causing a

nuisance or badly affected by Dutch elm disease.

In addition to these exceptions, an owner of woodland may, without licence, cut a total of 5m³, and sell up to 2m³ of that amount, in any three-month period.

Applications for a felling licence should be made to the local officer of the Forestry Authority, with a map of the property marked to show the limits of the area where felling is proposed. The application should say how many trees are to be felled, their species and the estimated volume. The Forestry Commission's free leaflet 'Control of Tree Felling' is available from all their offices, giving guidance and the latest rules.

A woodland owner who has a current management plan for the woods should have no difficulty in receiving a licence. When granted it will normally stipulate as a condition that the area should be promptly and effectively regenerated. The usual expectation is that broadleaved woodland will be replaced with broadleaves, and mixed broadleaved-conifer woods will be replaced with mixed woods. Natural regeneration is usually fully acceptable for a grant under the Woodland Grant Scheme.

TREE PRESERVATION ORDERS

Tree Preservation Orders (TPOs) are issued by local authorities, making it necessary for the owner or occupier of land to obtain the prior consent of that authority before any lopping, topping or felling is carried out. The Orders are made under the Town and Country Planning (TPO) Regulations and the owner of the trees affected must be informed by the Council making the Order, with the trees clearly identified on a map. The seller of land must tell a buyer that a TPO has been made in respect of trees on the property. Do not be involved in cutting a tree covered by an order; in one recent case an English court imposed a fine of £16,000 and costs, after the unlawful felling of three trees, a birch and two sycamores.

Whereas the Council's intention in imposing a TPO is to conserve a tree and to protect amenities, they may achieve the opposite in the longer term. Protected trees may eventually die of neglect and old age with no arrangements for their replacement, a situation now well understood by Planning Officers. TPOs are useful for a short time in stopping irresponsible cutting of trees, but they do nothing to ensure good management and sustention.

An owner who has woods covered by a TPO and who intends to practise sustainable management should apply through the Planning Officer of the Council which imposed the TPO to have the Order lifted in order to allow positive and conservative management. A Planning Officer should be willing to lift or vary the Order if he is persuaded the owner's proposals will conserve the woodland and thus achieve the Order's long-term objective.

The most troublesome TPO issues are usually those involving very old prominent trees which are landmarks loved by the public and standing in or immediately beside public places, main roads, etc. When these trees become potentially dangerous the question of legal responsibility for retaining them arises, since their owner, who carries the responsibility for damage they may do, is prevented by the Council from felling them. Heavy premiums may be required to provide special public liability insurance to indemnify the owner against damage claims and owners should discuss with the Council who should pay the premiums.

If the trees to be felled are covered by a TPO but are one of the exceptions from the felling licence requirements, an application for consent to fell should be made directly to the Council planning authority. If, however, the trees are of a class requiring a felling licence, application should be made to the Forestry Commission who will forward it, with their comments, to the Council.

PLANTING CLEARANCE

Although the obtaining of clearance for afforestation proposals is not a matter of law, it is appropriate to include it here.

Since the environmental impact of new planting can be great, the Forestry Authority informs some statutory bodies about planting proposals and seeks their advice before agreeing or refusing applications for planting grants. In effect, the process amounts to obtaining clearance for planting. The bodies approached include the government agricultural departments, water authorities, the local authority, the nature conservation agencies, the Deer Commission in the Scottish Highlands, and so on.

There is provision for large afforestation proposals to be submitted, in case of unresolved disagreement, to regional advisory committees, but this is very unlikely to concern small woodlands. Any large scheme likely to have a significant impact on the environment may be required by the Forestry Authority to have an environmental assessment submitted in the first instance, the cost of which falls on the applicant.

UNSAFE TREES

It may happen that a tree which was apparently healthy and strong falls without warning or is blown down in a gale, causing severe damage to people and property. The owner of the tree would not be held responsible in law for the damage if the tree's failure could not reasonably have been foreseen or prevented. On the other hand, if the tree had obvious signs of decay or weakness, the owner should expect to be held financially liable for the damage caused by his dangerous tree and to be successfully sued. In the event of the tree falling on a public path and causing injuries, or across a boundary to cause a road traffic accident, the damages could be very large. Consequently, there is reason for the owner of trees to look carefully at trees near boundaries. There is no need for specialised knowledge in such an inspection; what is required for the judgement of the court is the sharp eye of a sensible person to decide what could reasonably be foreseen as a danger.

In looking at trees, especially near boundaries and places where the public may have access, look particularly for leaning trees and those with roots that heave up in the wind, those with large dead branches or with thin crowns, shedding leaves before autumn, those with bracket fungi growing from the stem or toadstools round the base, and those which are hollow, have pockets of rot in them or have areas of dead bark. Trees which have had roots disturbed in trench-cutting, the soil compacted by traffic, the soil level altered in landscaping operations or the bole surrounded with tar and concrete are especially suspect.

Well-managed woods in which trees are harvested at normal maturity seldom give cause for any anxiety. Neglected trees left to waste into senility are the dangerous ones.

NEIGHBOURS

Trees on boundaries, even when healthy and no physical hazard, may be the cause of contention between neighbours. The problem may be one of shading, dripping or sending roots into a neighbour's land.

In law, a landowner has no right to the light and air space or to the root space in the earth beyond the boundary line of his property. If a dispute arises concerning a tree near a boundary, the sequence of events to reduce the nuisance might be as follows. The person suffering nuisance should specify it (overhanging branches, drip, roots invading tile drains, undermining a wall, etc.) and ask the owner of the tree to act in order to reduce it. The owner of the tree may then cut the branches or roots, or he may ask the neighbour to cut them where they cross the boundary, at the tree-owner's expense.

If the owner of the tree does not act reasonably promptly to put the matter right, the neighbour may take action himself, first giving notice that he intends to cut the offending branches or roots, and taking care he does not damage the tree beyond the boundary, since that is not his property; he may also sue for damages if actual damage has been caused, for instance by roots disturbing the foundations of walls.

The owner of woodland has special responsibility in law to neighbours in respect of water. Cultivation and drainage must be done so that neighbours' water rights and fisheries are not harmed and special care must be taken when timber trees are being harvested that streams are not affected by silt and felling debris; this type of damage is likely to affect fish breeding by covering eggs or breeding gravel beds with mud.

If the owner of trees with poisonous foliage, such as yew, allows them to grow beyond the boundary where they are eaten by a neighbour's stock, the tree's owner will be liable for damages. And you may be sure the dead animal will be a pedigree bull or a thoroughbred racehorse; that is Macpherson's Law.

FURTHER READING

Forestry Commission (1985), *External Signs of Decay in Trees*, Forestry Commission Arboricultural Leaflet 1, HMSO, London
Gordon, W.A. (1955), *The Law of Forestry*, The Colonial Office, HMSO, London

Epilogue: What Is It For?   19

The forester's task is not to exploit the forest but to ensure its perpetuation.

In the mid-nineteenth century the thriving forest industry in the whole of the British Isles virtually stopped when superb-quality timber was imported at unmatchably low prices from the destruction of virgin forests in North America and elsewhere. These timbers carried no costs of growing or replacement; home sawmills could not compete and, as they stopped buying logs, so felling in the woods had to stop, and planting stopped. The woodlands were no longer a sustained system. The long tradition of multi-purpose management was broken and the technical skills it required were steadily lost.

The decline of woodlands in Britain was then hastened by the break-up of many country estates which, at their best, had left farming to tenant farmers and the woods in the care of the landowner and a skilled forestry squad. With the change of farms to owner-occupation, the woodlands, already in poor condition, came into the hands of farmers who had neither experience of woodland skills nor, in the prevailing timber market, any incentive to manage them positively.

The poverty of British wood-growing resources and the consequent vulnerability of the country to timber shortage in the First World War induced the government in 1919 to adopt a new policy of state investment in forestry. The great afforestation drive which followed, largely on bare upland, left largely unaffected the relict small woods on lowland farms and former estate policies. Many linger on, neglected, having been further depleted of any quality timber by the enforced and often selective exploitation in the Second World War.

Even with the peak of the twentieth-century afforestation now coming into timber production, the United Kingdom currently imports four-fifths of the wood products it consumes, at a net cost which exceeds £4,000 million annually. Demand for wood provides a huge market.

The long period in which the world's wood markets have been driven by the destructive exploitation of virgin forests is virtually at an end. Many countries which were massive exporters of wood have now exhausted their forests and have become heavy importers. Increasingly the price of foreign timber includes the realistic costs of the suppliers running their forests as sustainable resources. As the supply of tropical hardwoods diminishes, so the prospects improve for home-grown timber of good quality. There was never a better opportunity to begin the restoration of woodlands. The fact remains that the time required for growing large hardwoods is long, so the sooner one begins the better.

The decades of conifer afforestation in the uplands were a rational response to the lack of industrial wood resources, since industry's bulk demand is for softwoods. One extreme reaction now is to suggest switching to the planting of strictly indigenous species, in effect pushing the pendulum from plantings for timber, adversely criticised as serving too narrow a purpose, to the creation of fragments of wild wood which would be unlikely to produce any useful timber at all and are open to the same criticism, albeit for a different single purpose. Such woodlands, because of the cultural practices adopted, are in danger of being unable to pay for their own maintenance in the decades ahead and of becoming the derelict woods of the future, always prey to someone who sees more profit in the land being otherwise used. Internationally and locally the expectation and requirement is for sustained development of resources. If that policy is to achieve more than lip-service with regard to woodlands in Britain, it requires the reapplication of multi-purpose management.

The huge number of run-down small woodlands in the lowlands of Britain urgently need sensitive care to ensure their perpetuation. The great challenge is how to make them at the same time productive, ecologically sensitive, visually pleasing and compatible with neighbouring management, perhaps to be financially viable in the short term, but surely to hold the promise of high value and benefit in the future.

Take the opportunity to plant trees and sustain woods now and 100 or 200 years hence someone may think well of you, even if they will not know your name; there are not many things we may do in this world to achieve that.

Glossary 1

Terms	
Adventitious	Arising from an unusual place, e.g. buds developing on the main stem instead of in the axil of a leaf.
Afforestation	Strictly, the first planting of land previously without trees; subsequent planting of woodland is *reforestation* or regeneration.
Angiosperms	Flowering plants having seeds contained in a 'box' (carpel or ovary, with stigma); comprises both Dicotyledones (two seed leaves, including all broadleaved trees) and Monocotyledones (one seed leaf: grasses, bamboos and palms).
Arboretum	A collection of many species of trees, grown as single specimens or small groups, for scientific study or as a tree garden.
Arboriculture	The growing of trees as individuals, as distinct from *silviculture*, the husbandry of woods.
Bast	The inner living bark which conducts sugary sap down from leaves to roots; after the cells die it becomes bark.
Beating-up	The replacement of losses in a young plantation, usually in the year after first planting.
Bole	Tree's main stem or trunk.
Bract	Leaf-like structures at base of flower or cone scale.
Brashing	The removal of dead side branches from standing pole-stage conifers, to a height of about 2m, to allow better access and sometimes as a fire precaution.
Breast height	The standard height for measuring the (nominally) basal diameter (or girth) of standing trees, now 1.3m above ground level.
Broadleaved tree	A tree of the natural order Dicotyledones, most having broad, flat leaves (in Britain mostly also deciduous); a *hardwood*, e.g. oak, birch, ash.
Brown earth	A fertile soil type in which the humus is well incorporated into the soil profile, generally by the action of abundant earthworms; typically base-rich, neutral or only moderately acid reaction, with no sharp divisions between litter and lower soil horizons.
Burr	A solid woody growth on a tree bole, usually formed as a result of extraordinary growth of dormant buds or stimulation of the cambium by insects or viruses; often has decorative wood.
Butt	The larger, basal end of a tree trunk or log. The *butt log* is the lowest cut from a tree trunk.

Callus	Tissue of thin-walled cells developing at the site of an injury to heal and occlude it; used for plant-breeding by tissue culture.
Cambium	Layer of actively dividing cells over the whole surface of the living tree between bast and wood, producing wood cells (xylem) on its inner side and bast (phloem, becoming bark when it dies) on the outer side.
Canker	Open wound caused by fungus or bacterium.
Canopy	The whole assembly of tree crowns in a wood, designed to intercept sunlight for photosynthesis; there may be a single canopy, or two or more.
Cant-hook	A tool for turning or rolling logs.
Catkin	A cluster of small wind-pollinated flowers.
Chlorophyll	The green pigments involved in photosynthesis.
Cleaning	The operation in which weed trees, climbers and badly formed specimens of the crop species are removed from a stand of young trees, generally at the thicket stage.
Clone	A strain of trees propagated vegetatively from a single individual (by cuttings, tissue culture, etc.); to propagate thus.
Collar	The point on the tree stem where the root meets the shoot; the natural ground level.
Compartment	An area of woodland defined for management convenience of description, record, etc.; the basic and permanent unit of management in a forest.
Compound	Composed of many leaflets etc.
Compression wood	The reaction wood (qv) of conifers, formed on the underside of horizontal branches and leaning trunks, characterised by denser than normal structure and higher lignin content than normal wood. It makes unsatisfactory timber (twisting) and pulp.
Conifer	A tree of the natural order Coniferae, distinguished by seeds borne unenclosed on the scales of cones; the leaves are mostly needle-like or scale-like and most are evergreen.
Coppice	To cut back broadleaved trees so that they may sprout again from the stump or rootstock; an area of woodland so managed, traditionally cut on a rotation between eight and 35 years, very recently as short as three.
Coppice-with-standards	An area of woodland managed as coppice but with scattered trees not cut on the coppice rotation but grown as large timber.
Cord	A measure of volume for stacked branchwood and small stemwood; 128 cubic feet.
Cotyledon	The first seed leaf of a seedling; Dicotyledones have two; Monocotyledones (grasses, bamboos etc.) have one; Coniferae have two to 15 or more.
Crown	The whole branch system and foliage of a tree.
Deciduous	Shedding all its leaves during part of the year, normally in winter in northern Europe.
Diffuse-porous	Wood with the vessels spread evenly throughout the spring and later wood (e.g. birch and beech).

Direct sowing	Sowing tree seed directly in their intended permanent place in the forest, rather than in a nursery.
Double leader	A forked tree; a plant with two stems of equal vigour and size.
Epicormic	Arising from the outer layer of the stem.
Even-aged stand	A compartment or wood composed of trees of approximately the same age.
Evergreen tree	Having leaves throughout the year; some leaves are grown annually and some shed annually, the life of a given leaf being from one to about five years, exceptionally longer.
Group clear-felling	A system of forest management involving cutting rather large groups for regeneration, the age range of the stands being considerably less than the rotation.
Group selection	A system of forest management involving cutting rather small groups for regeneration, the age range of the groups covering the full rotation or production period of the trees, thus providing perpetual woodland cover.
Gymnosperm	Plants with seeds not enclosed in a box or carpel, the pollen germinating directly on the surface of the ovule; comprises mostly the conifers but also cycads and gnetums.
Hardwood	A broadleaved tree, or the wood from it.
Heart-rot	The condition, arising from disease such as Fomes, in which the central core of the tree bole becomes rotten, perhaps hollow, for some distance up from ground level.
Heartwood	Wood in the inner part of the tree, transformed by chemical changes due to ageing and no longer carrying sap.
Heeling-in	Burying the roots of young trees in soil temporarily in order to keep them moist while awaiting planting in their final positions (to *sheugh-in* in Scots).
High forest	A stand, irrespective of actual height, in which the trees are derived from seedlings and are intended to continue so; cf. coppice.
Humus	Structureless organic matter (animal or vegetable in origin) in the soil.
Hybrid	Trees arising from the cross-breeding, artificial or natural, of two related species, usually within the same genus, occasionally not.
Ironpan	An impermeable layer in an acid, podzolised soil, formed by the deposition of iron (and perhaps manganese etc.) salts leached from upper layers; severely restricts rooting.
Juvenile wood	Low-density and weak wood formed in the tree's early years and thereafter within the live crown. Present in broadleaves but much more important in conifers.
Knot	The base of a branch buried as the stem-wood grows around it; normally denser, harder, more resinous than the surrounding wood (affecting wood-working and finishing). Knots decrease the bending strength of battens because the grain is at a large angle to the stem-wood grain. A *live knot* or *intergrown knot* results from the base of a live branch being grown into the stem, the wood fibres of the two being continuous. A *dead knot*

or *encased knot* results from the inclusion of a now-dead branch into the stem-wood, there being no continuity of fibres between them. A *powder knot* results from the inclusion of a rotting branch (especially common, and serious, in pines).

Layer	A plant propagated vegetatively by pegging down and partly burying in soil a living shoot of the parent, severed after rooting; to propagate thus.
Leader	The main upright shoot of a tree.
Lenticel	Corky breathing pore on young tree bark.
Light-demander	A tree species which is intolerant of overhead shade.
Line thinning	A system of thinning a plantation involving the removal of complete lines of trees.
Litter	The accumulation of undecomposed leaves etc. on the soil surface.
Maiden	A tree grown from seed and not cut back at any time.
Mast	Broadleaved tree seed crop, especially that of oak and beech; *mast year*, a year of abundant seed production.
Mature	Ready for cutting; at the end of the rotation for that particular crop.
Medullary ray	Band of living cells grown radially in wood, used in the living tree for food storage; often attractive in the timber, e.g. silver grain in oak.
Moder	A form of humus intermediate between mull and mor; it has an attractive 'earthy' smell.
Mor	Humus type, typical of heathland, greasy, black, acid and poor in micro-fauna.
Mull	Crumbly humus type, well incorporated in the mineral soil and about neutral pH, the litter being rapidly broken down by micro-fauna and earthworms.
Mycorrhiza	A root structure involving a growing together of a tree's fine roots and the threads of a fungus, to their mutual benefit; it enables the tree to derive nutrients directly from decomposing plant litter.
Natural regeneration	The regrowth of tree stands from self-sown seeds and suckers, without planting or artificial sowing.
Notching	Planting a tree in a narrow slit or a combination of slits cut into the ground.
Nurse	A tree grown principally to shelter a tree of another species which is potentially more valuable.
Pan	A hard layer in the soil, usually impenetrable by roots, caused by soil chemical action (iron or manganese) or by compaction.
Peavey	A tool for turning logs.
Peduncle	Flower stalk.
Petiole	The stalk of a leaf.
Photosynthesis	The chemical process in green plant tissues, driven by light energy, in which carbon dioxide in the air is reduced to carbohydrates (sugars etc.) and oxygen is released from water in the plant sap.
Pioneer	A tree naturally adapted to colonise bare land.

Pit-planting	Planting young trees in a prepared hole large enough to spread the full root system naturally.
Podzol	An acidic soil type displaying three sharply defined layers: a dark humus layer, a zone of bleached soil from which iron, manganese and aluminium have been leached, and a zone, usually dark coloured, beneath that where these elements have been redeposited; common on heathlands.
Pollard	To cut a tree bole 2m or more above ground level so that it will shoot again, identical to coppicing but the shoots out of reach of stock.
Pruning	The removal of side branches to increase the volume of knot-free timber and to improve the form of a tree by the removal of double leaders etc.
Pure stand	A stand of trees formed entirely or almost of one species.
Quarter-sawn	Timber sawn on (or near) the radius of the stem in order to show the figure of the medullary rays (i.e. the silver grain), as in oak, and to minimise distortion.
Reaction wood	Anatomically distinctive wood formed under tension or compression, tending to maintain the position of the growing branch or leaning stem. It makes unsatisfactory sawn timber or pulp (see *compression wood* and *tension wood*).
Resin-canal	Duct in conifer wood and leaves; lined with resin-producing cells and conducting resin.
Ride	A broad open track through a wood, not hard-metalled as a road.
Ring-porous	Wood with much larger and/or more numerous vessels in the early spring wood than in the later wood, so that the cross-section of the spring wood is seen as a ring of fine holes (e.g. deciduous oaks and sweet chestnut).
Rotation	The production period of a forest stand; the period from regeneration to final felling.
Sap	A spiked pole used for moving logs. Also the watery solution in the wood or bast.
Sapling	A young tree, larger than a seedling, smaller than a pole.
Sapwood	Outer wood carrying sap upward from roots to leaves.
Screef	To cut off a thin surface layer on the ground.
Selection forest	A forest in which every age class is represented on each unit area from seedling to maturity; it may be a stem-by-stem selection forest in which the different ages are intimately mixed, or a group-selection one, in which the ages are represented by small, even-aged groups.
Sessile	Stalkless.
Shade-bearer	A species which can tolerate some shade while still surviving and growing.
Silviculture	The art and science of growing trees in woodland.
Slab	The waste wood cut from the outside of a log when it is squared in the sawmill.
Softwood	A coniferous tree and the wood therefrom.
Spring wood	Soft, pale band of wood, mainly for conducting sap, laid down each year in spring.
Stand	A population of trees forming a unit for management.

Stomata	Minute breathing pores on the under-surface of the leaf.
Stool	The living stump of a cut tree, especially a broadleaved tree cut for coppice.
Store	To leave uncut trees in a coppice crop to grow on as timber; a tree so left.
Straining post	A heavy post at the change of direction of a fence, holding the stretched wires of the straight lengths.
Sucker	Young tree arising from the root of an older one.
Summer wood	Dense darker band of wood, structurally strong, laid down each year in summer.
Suppressed tree	A tree outgrown by its neighbours and now entirely under their crowns.
Tension wood	The *reaction wood* (qv) of broadleaves, found on the upper side of horizontal branches and leaning stems, characterised by lower than normal lignin content. It makes unsatisfactory sawn timber and pulp.
Tracheid	A strong elongated wood cell with pointed ends and pits in the side walls to transmit water; the main structural element in conifer wood.
Transpiration	The loss of water by evaporation, mainly through the stomata of the leaves.
Transplant	A seedling tree germinated in a seedbed and then moved once (or twice) to another nursery bed before lifting for final planting.
Undercutting	The process of deliberately cutting the roots of tree plants in the nursery, without lifting them, as a substitute for transplanting, with the object of inducing the growth of fine roots.
Underplanting	Planting trees below an existing stand, to improve it or eventually to replace it.
Underwood	Coppice or shrub species existing beneath a main stand of a forest.
Vessel	In broadleaved wood, a long unbranched tube, maybe several metres, for conducting water; formed by the breakdown of the end-walls of adjoining cylindrical cells.
Wane	The curved outside surface of the trunk sometimes left on the edges of sawn squared timber.
Whip	A weak spindly tree liable to damage its neighbours by threshing them in a wind.
Wolf	A coarse heavily branched tree in a young stand, undesirable because it damages many (potentially better) neighbours.
Yield table	A table, based on research plots, used to estimate the yield of timber, thinnings, etc. from a stand of a given species, under defined conditions, at various ages.

Glossary 2

Scientific names and authorities of trees with the English names used in the text.

ALDER
	Common	Alnus glutinosa (L.) Gaertner
	Cordate or Italian	Alnus cordata (Lois.) Duby
	Grey	Alnus incana (L.) Moench
	Red	Alnus rubra Bong.

APPLE
| | Siberian Crab | Malus baccata (L.) Borkh. |
| | Wild | Malus sylvestris (L.) Miller |

ASH   Fraxinus excelsior L.
ASPEN   Populus tremula L.
BEECH   Fagus sylvatica L.
BIRCH
| | Downy | Betula pubescens Ehrh. |
| | Silver | Betula pendula Roth |

BOX   Buxus sempervirens L.
BUCKTHORN
| | Common | Rhamnus cathartica L. |
| | Sea | Hippophae rhamnoides L. |

CEDAR
	Atlas	Cedrus atlantica (Endl.) Carrière
	Lebanon	Cedrus libani A. Richard
	Western Red	Thuja plicata Donn *ex* D.Don

CHERRY
	Bird	Prunus padus L.
	Plum	Prunus cerasifera Ehrh.
	Wild (Gean)	Prunus avium L.

CHESTNUT
	Horse	Aesculus hippocastanum L.
	Red	Aesculus x carnea Zeyher
	Sweet	Castanea sativa Miller

COAST REDWOOD   Sequoia sempervirens (D.Don) Endlicher
CYPRESS
| | Lawson | Chamaecyparis lawsoniana (A.Murray) Parlatore |

	Leyland	xCupressocyparis leylandii (Jacks. & Dallim.) Dallim.
	Monterey	Cupressus macrocarpa Hartweg *ex* Gordon
	Nootka	Chamaecyparis nootkatensis (D.Don) Spach
DEODAR		Cedrus deodara (Roxburgh *ex* D.Don) Don
ELM		
	English	Ulmus procera Salisb.
	Wych	Ulmus glabra Hudson
FIR		
	Caucasian	Abies nordmanniana (Steven) Spach
	Douglas	Pseudotsuga menziesii (Mirbel) Franco
	European Silver	Abies alba Miller
	Grand	Abies grandis (Douglas *ex* D.Don) Lindley
	Noble	Abies procera Rehder
GEAN		Prunus avium L.
GUELDER ROSE		Viburnum opulus L.
HAWTHORN		Crataegus monogyna Jacq.
HAZEL		Corylus avellana L.
HEMLOCK		
	Western	Tsuga heterophylla (Raf.) Sargent
HOLLY		Ilex aquifolium L.
HORNBEAM		Carpinus betulus L.
JUNIPER		Juniperus communis L.
LARCH		
	European	Larix decidua Miller
	Hybrid	Larix x eurolepis A. Henry
	Japanese	Larix kaempferi (Lindley) Carrière
LIME		
	Common	Tilia x vulgaris Hayne
	Large-leaved	Tilia platyphyllos Scop.
	Small-leaved	Tilia cordata Miller
LOCUST TREE		Robinia pseudacacia L.
MAPLE		
	Field	Acer campestre L.
	Norway	Acer platanoides L.
MIDLAND THORN		Crataegus laevigata (Poiret) DC.
OAK		
	Holm	Quercus ilex L.
	Pedunculate	Quercus robur L.
	Red	Quercus rubra L.
	Sessile	Quercus petraea (Mattuschka) Lieblein
	Turkey	Quercus cerris L.
OSIER		Salix viminalis L.
PEAR		Pyrus communis L.

PINE	
Austrian	Pinus nigra Arnold *var.* nigra
Corsican	Pinus nigra Arnold subsp. laricio Maire
Lodgepole	Pinus contorta Douglas *ex* Loudon
Maritime	Pinus pinaster Aiton
Radiata	Pinus radiata D.Don
Scots	Pinus sylvestris L.
Western yellow	Pinus ponderosa Douglas
Weymouth	Pinus strobus L.
PLANE	
Buttonwood	Platanus occidentalis L.
London	Platanus acerifolia (Aiton) Willd.
Oriental	Platanus orientalis L.
POPLAR	
Black	Populus nigra L. subsp. betulifolia (Pursh) W.Wettst.
Grey	Populus x canescens (Aiton) Smith
White	Populus alba L.
ROWAN	Sorbus aucuparia L.
SPRUCE	
Norway	Picea abies (L.) Karsten
Sitka	Picea sitchensis (Bong.) Carrière
STRAWBERRY TREE	Arbutus unedo L.
SYCAMORE	Acer pseudoplatanus L.
WALNUT	
European	Juglans regia L.
Black	Juglans nigra L.
WAYFARING TREE	Viburnum lantana L.
WELLINGTONIA	Sequoiadendron giganteum (Lindley) Buchholz
WESTERN RED CEDAR	Thuja plicata Donn *ex* D.Don
WHITEBEAM	Sorbus aria (L.) Crantz
Arran	Sorbus arranensis Hedlund
Finnish	Sorbus hybrida L.
Swedish	Sorbus intermedia (Ehrh.) Pers.
WILD SERVICE TREE	Sorbus torminalis (L.) Crantz
WILLOW	
Bay	Salix pentandra L.
Crack	Salix fragilis L.
Goat	Salix caprea L.
Osier	Salix viminalis L.
White	Salix alba L.
YEW	Taxus baccata L.

Appendix 1

WOODLAND GRANT SCHEME

Grants are paid for the following (the rates are subject to periodic review):

Establishment by planting

Flat rate per ha (£700 in 1997) for any size of conifer planting and almost double the conifer rate for broadleaves up to 10ha; 50 per cent above the conifer rate for broadleaves over 10ha. Native pinewood planting is paid at the broadleaved rate (north of the Forth-Clyde only). Grants are paid in two instalments, 70 per cent after planting and the balance at year five. Normal stocking density is 2,250 trees per ha (2.1m apart) but a lower density of 1,100 per ha may be accepted for some woodlands of native species where no timber is expected etc., and for poplar planting.

Community Woodlands Supplement

The supplement (£950 per ha in 1997) may be paid to encourage the planting of new woodlands close to urban areas, where the owner allows the woodland to be used for informal public recreation (i.e. walking etc.). There is a strict quota for each urban population. The full amount is paid after planting.

Better Land Supplement

Flat rate (£600 per ha in 1997) paid for new planting on arable or improved grassland (except grassland not used in farming). The full amount is paid after planting or successful natural regeneration.

Short-Rotation Coppice

This system is grant-aided for a five-year trial period from 1995, limited to 1,250ha per annum on set-aside land and 1,000ha per annum on non-set-aside land, at flat rates of £400 per ha on set-aside and £600 per ha on non-set-aside land. The full amount is paid after planting; other supplementary grants are not payable with this grant.

Restocking

Flat-rate grants of £325 per ha for conifers (1997) and £525 per ha for broadleaves, paid in full after planting.

Restocking by natural regeneration

Flat rates per ha at the same rates as restocking above. Grants are paid in two instalments: a discretionary payment for half agreed costs of soil preparation etc. and the balance when adequate stocking is achieved.

Annual Management Grant

This grant (£35 per ha in 1997) is available for any age of woodland for agreed work to enhance its value or bring it up to currently accepted standards or to allow public access.

Woodland Improvement Grant

These are discretionary capital payments at 50

per cent of the agreed cost of work in existing woodland to achieve significant public and environmental benefits (opening for public recreation, restoring neglected woodlands and promoting biodiversity).

Livestock Exclusion Annual Payments

These are annual payments for ten years (£80 per ha per annum in 1997) to compensate for the loss of grazing and shelter where woodland is closed to farm stock, where regeneration and protection are a priority; available in 'Less Favoured Areas' or 'Environmentally Sensitive Areas' only. The payments are treated as agricultural income for tax purposes.

Locational Supplement

An additional incentive for the planting of new woodland in specially targeted areas (e.g. the Central Scotland Forest). The supplement of £600 per ha (in 1997) is paid after planting.

Tender Schemes

The National Forest Tender Scheme is a pilot aimed at stimulating new planting in special targeted areas, initially the National Forest in the English Midlands.

Farm Woodland Premium Scheme

These grants are designed to encourage the planting of new woodlands on farms, both for wood production and to enhance the environment. They are administered by the Agricultural Departments (MAFF, SOAFD and WOAD), being annual payments in addition to the Woodland Grant Scheme, paid for selected projects. No work must be undertaken until approval has been obtained from the relevant agricultural department. Payments, at rates between £60 and £300 per ha per annum (1997), are made for 15 years where the planting is more than half broadleaved and for ten years where it is less than half; the minimum area is 1ha and the maximum 200ha (or 40ha of unimproved grassland). Payments are regarded as farm income for tax purposes.

For all the above grants and schemes the first contact should be the Forestry Authority local Conservancy office, which will supply an applicant's pack for WGS and FWPS Grants, although the Farm Woodland Premium Scheme is administered by the country Agricultural Departments. Other financial assistance may be available for tree-planting or woodland improvement, as discretionary grants, from the nature conservation agencies, local authorities, local enterprise bodies, etc., to whom application should be made directly.

Appendix 2

LIST OF ADDRESSES

Arboricultural Advisory and Information
  Service
Alice Holt Lodge
Wrecclesham
Farnham
Surrey GU10 4LH

Arboricultural Association
Ampfield House
Romsey
Hampshire SO51 9PA

British Association for Shooting and
  Conservation
Marford Mill
Rosset
Wrexham
Clwyd LL12 0HL
and
Trochry
by Dunkeld
Perthshire PH8 0DY

British Christmas-Tree Growers' Association
12 Lauriston Road
Wimbledon
London SW19 4TQ

British Horse Loggers Specialist Group
Forest Contracting Association
Dalfling
Blairduff
Inverurie
Aberdeenshire AB51 5LA

British Timber Merchants' Association
Stocking Lane
Hughenden Valley
High Wycombe
Bucks HP14 4JZ

British Trust for Conservation Volunteers
36 St Mary's Street
Wallingford
Oxon OX10 0EU

Charcoal Association
Eastern Cottage
Main Road
Toft
Bourne
Lincs PE10 0JT

Coillte Teoranta
Leeson Lane
Dublin 2
Ireland
Irish Forestry Board

Community Development Foundation
60 Highbury Grove
London N5 2AG

Coppice Association
Eastern Cottage
Main Road
Toft
Bourne
Lincolnshire PE10 0JT

Country Landowners Association
16 Belgrave Square
London SW1X 8PQ

Countryside Commission
John Dower House
Crescent Place
Cheltenham
Gloucestershire GL50 3RA
Official agency for countryside affairs in England

Countryside Commission Community
   Forest Unit
4th Floor
71 Kingsway
London WC2B 6ST

Countryside Council for Wales
Plas Penrhos
Ffordd Penrhos
Bangor
Gwynedd LL57 2QL
Official agency for countryside affairs and nature conservation in Wales

English Nature
Northminster House
Peterborough PE1 1UA
Official agency for nature conservation in England

English Partnerships
16–18 Old Queen Street
London SW1H 9HP
Administers Derelict Land Grant in England

Environment and Heritage Service (NI)
(DOE)
Commonwealth House
35 Castle Street
Belfast, BT1 1GU
Official agency for nature conservation in Northern Ireland

Farming and Wildlife Advisory Trust
National Agricultural Centre
Stoneleigh
Kenilworth
Warwickshire CV8 2RX
(Also county offices listed under
Farming and Wildlife Advisory Group
in telephone directories)

Forest and Wildlife Service
Leeson Lane
Dublin 2
Ireland

Forest Service (NI)
Department of Agriculture for Northern
   Ireland
Dundonald House
Upper Newtownards Road
Belfast BT4 3SB
Government department for agriculture and forestry in Northern Ireland

Forestry and Arboricultural Safety and
   Training Council
231 Corstorphine Road
Edinburgh EH12 7AT
Safety leaflets and courses

Forestry Authority England
Great Eastern House
Tenison Road
Cambridge CB1 2DU
Official agency for forestry authority in
England: felling licences and planting grants
through local offices

Forestry Authority Scotland
Portcullis House
21 India Street
Glasgow G2 4PL
Official agency for forestry authority in
Scotland: felling licences and planting grants
through local offices

Forestry Authority Wales
North Road
Aberystwyth
Dyfed SY23 2EF
Official agency for forestry authority in Wales:
felling licences and planting grants through
local offices

Forestry Commission Headquarters
231 Corstorphine Road
Edinburgh EH12 7AT
National (GB) department for forestry
(England, Scotland and Wales) including
Forestry Authority (three agencies) and Forest
Enterprise

Forestry Contracting Association
Dalfling
Blairduff
Inverurie
Aberdeenshire AB51 5LA

Forestry Industry Council of Great Britain
Golden Cross House
3–8 Duncannon Street
London WC2N 4JF

Forestry Research
Alice Holt Lodge
Wrecclesham
Farnham
Surrey GU10 4LH
and
Northern Research Station
Roslin
Midlothian EH25 9SY

Game Conservancy Ltd
Burgate Manor
Fordingbridge
Hants SP6 1EF

Health and Safety Executive
Library and Information Services
Broad Lane
Sheffield S3 7HQ

Horticultural Trades Association
19 High Street
Theale
Reading
Berkshire RG7 5AH

Institute of Chartered Foresters
7A St Colme Street
Edinburgh EH3 6AA
Chartered professional forestry institute with
register of approved forestry consultants

National Small Woods Association
3 Perkins Beach Dingle
Stiperstones
Shropshire SY5 0PF

Northern Ireland Forest Service
Department of Agriculture
Dundonald House
Upper Newtownards Road
Belfast BT4 3SB

Royal Forestry Society of England, Wales and
Northern Ireland
102 High Street
Tring
Herts HP23 4AF
*Forestry society for England, Wales and
Northern Ireland; general forestry and
woodland interest; quarterly journal*

Royal Scottish Forestry Society
62 Queen Street
Edinburgh EH2 4NA
*Forestry society for Scotland; general forestry
and woodland interest; quarterly journal*

Scottish Conservation Projects
Balallan House
24 Allan Park
Stirling FK8 2QG

Scottish Forestry Trust
5 Dublin Street Lane South
Edinburgh EH1 3PX
*Charitable trust for forestry education, training
and research*

Scottish Landowners' Federation
25 Maritime Street
Edinburgh EH6 5PW

Scottish Natural Heritage
12 Hope Terrace
Edinburgh EH9 2AS
*Official agency for countryside affairs and
nature conservation in Scotland*

Scottish Wildlife Trust
Cramond House
16 Cramond Glebe Road
Edinburgh EH4 6NS

Society of Irish Foresters
2 Lower Kilmacud Road
Stillorgan
Co. Dublin
Ireland

Timber Growers' Association Ltd
5 Dublin Street Lane South
Edinburgh EH1 3PX

Tree Council
51 Catherine Place
London SW1E 6DY
*Charity to improve environment by planting
and conserving trees and woods throughout the
UK*

Welsh Development Agency
Pearl Assurance House
Greyfriars Road
Cardiff CF1 3XX
Administers Derelict Land Grant in Wales

The Wildlife Trust
The Green
Witham Park
Lincoln LN5 7JR

The Woodland Trust
Autumn Park
Dysart Road
Grantham
Lincs NG31 6LL

The Woodland Trust Scottish HQ
Glenruthven Mill
Abbey Road
Auchterarder PH3 1DP

Index